Chemistry with Vernier

Chemistry Experiments Using Vernier Sensors

Dan D. Holmquist
Jack Randall
Donald L. Volz

Measure. Analyze. Learn.™

Vernier Software & Technology
13979 S.W. Millikan Way • Beaverton, OR 97005-2886
Toll Free (888) 837-6437 • (503) 277-2299 • FAX (503) 277-2440
info@vernier.com • www.vernier.com

Chemistry with Vernier

Chemistry Experiments Using Vernier Sensors

Published by
Vernier Software & Technology
13979 SW Millikan Way
Beaverton, OR 97005-2886
(888) 837-6437
(503) 277-2299
FAX (503) 277-2440
info@vernier.com
www.vernier.com

ISBN 978-1-929075-41-6
First Edition
Third Printing
Printed in the United States of America

About the Authors

Dan D. Holmquist earned his B.A. in Premedical Science and his M.S.T. in Chemistry from the University of Montana. His first teaching positions were at Livingston Junior High School and Park Senior High School, in Livingston, Montana. During the sixteen years of his employment with the Department of Defense Dependents Schools, he was the chemistry mainstay at Frankfurt American High School, located in Frankfurt, Germany. He taught high school chemistry during nineteen of his twenty-two years in teaching, including General Chemistry, Honors Chemistry, ChemCom, and AP Chemistry. Holmquist has taught in two Woodrow Wilson Outreach Workshops. He has also conducted numerous workshops for the Department of Defense Dependents Schools and the European Science Teachers Association. Two highlights of his teaching career include his distinction as a Woodrow Wilson Dreyfus Fellow in Chemistry in 1986, and his selection as a Presidential Awardee for Excellence in Science Teaching in 1987. He is currently working as the chemistry specialist at Vernier Software, where he still enjoys working with teachers to implement computer usage in the classroom. In addition to this book, he is co-author of the book *Water Quality with Vernier*.

Jack Randall earned his B.A. in Chemistry and Mathematics from Olivet College and his M.Ed. from the University of Arizona. He taught chemistry at Interlochen Arts Academy, Michigan. Jack's teaching experience ranges from grades four through community college, and he cites bilingual education and remedial education in a thirteen-year career in the classroom. Before moving to Interlochen, Jack spent six years working as a technical services and development chemist with Dow Chemical Company. He has a wide range of experience with the use of graphing calculators in science curriculum. Jack has conducted workshops on the use of graphing calculators in data collection and analysis in chemistry. He has also conducted numerous seminars on the use of the CBL and CBL 2 for high school, middle school, and upper elementary school teachers and students. Jack worked with Texas Instruments during the 1993-4 school year to field test the prototype to the CBL. He is currently working as the chemistry specialist at Vernier Software, where he still enjoys working with teachers to implement calculator usage in the classroom

Donald L. Volz received his B.S. in Chemistry and Mathematics and his M.S. in Chemistry Teaching from Purdue University. He taught at Fontana High School in California before taking his first teaching position with the Department of Defense Dependents Schools in Japan. He subsequently transferred within DoDDS to Mannheim, Germany, where he served as a chemistry and physical science teacher and as science department chairperson at Mannheim American High School before retiring. He has taught several graduate level courses on the subject of using technology in the science lab, and he has conducted numerous science teacher workshops for the Department of Defense Dependents Schools-Germany and the European Science Teachers Association. He is a Dreyfus Master Teacher and a 1991 Presidential Awardee for Excellence in Science Teaching. In addition to this book, he is coauthor of *Physical Science with Vernier, Middle School Science with Vernier, Investigating Environmental Science through Inquiry,* and *Investigating Chemistry through Inquiry,* also published by Vernier Software & Technology.

Proper safety precautions must be taken to protect teachers and students during experiments described herein. Neither the authors nor the publisher assumes responsibility or liability for the use of material described in this publication. It cannot be assumed that all safety warnings and precautions are included.

Contents

Experiments

Appendices

Sensors Used in Experiments

#	Experiment	Temperature	Gas Pressure	pH	Voltage	Conductivity	Colorimeter or Spectrometer
1	Endothermic and Exothermic Reactions	1					
2	Freezing and Melting of Water	1					
3	Another Look at Freezing Temperature	2					
4	Heat of Fusion of Ice	1					
5	Find the Relationship: An Exercise in Graphing Analysis						
6	Boyle's Law: Pressure-Volume Relationship in Gases		1				
7	Pressure-Temperature Relationship in Gases	1	1				
8	Fractional Distillation	1					
9	Evaporation and Intermolecular Attractions	2					
10	Vapor Pressure of Liquids	1	1				
11	Determining the Concentration of a Solution: Beer's Law						1
12	Effect of Temperature on Solubility of a Salt	1					
13	Properties of Solutions: Electrolytes and Non-Electrolytes					1	
14	Conductivity of Solutions: The Effect of Concentration					1	
15	Using Freezing Point Depression to Find Molecular Weight	1					
16	Energy Content of Foods	1					
17	Energy Content of Fuels	1					
18	Additivity of Heats of Reaction: Hess's Law	1					
19	Heat of Combustion: Magnesium	1					
20	Chemical Equilibrium: Finding a Constant, K_c						1
21	Household Acids and Bases			1			
22	Acid Rain			1			
23	Titration Curves of Strong and Weak Acids and Bases			1			
24	Acid-Base Titration			1			
25	Titration of Diprotic Acid: Identifying an Unknown			1			
26	Using Conductivity to Find an Equivalence Point					1	
27	Acid Dissociation Constant, K_a			1			
28	Establishing a Table of Reduction Potentials: Micro-Voltaic				1		
29	Lead Storage Batteries				1		
30	Rate Law Determination of the Crystal Violet Reaction						1
31	Timed-Release Vitamin C Tablets			1			
32	The Buffer in Lemonade			1			
33	Determining the Free Chlorine Content of Swimming Pool						1
34	Determining the Quantity of Iron in a Vitamin Tablet						1
35	Determining the Phosphoric Acid Content in Soft Drinks			1			
36	Microscale Acid-Base Titration			1			

Preface

This book contains thirty-six chemistry experiments using Vernier LabQuest, Vernier LabPro, Texas Instruments CBL 2, Vernier Go!Link, or Vernier EasyLink for collecting data. Your students can collect, display, print, graph, and analyze data in well over fifty percent of the experiments in a chemistry course. High-quality Vernier sensors will result in more accurate temperature, pH, voltage, colorimetric, conductivity, and pressure data measurements in the chemistry lab. Your students will perform many new experiments with measurements not previously obtainable in the classroom. Investigations formerly done with devices such as thermometers, pH meters, and manometers are greatly enhanced when electronic sensing is used.

You will discover many advantages to using probeware for data collection. Using Vernier probes opens up a large number of new chemistry investigations—the 36 experiments we have included here are only a small fraction of the many possibilities.

You will find a wide range of experiments in this book. Whether your chemistry class is high school or college, Advanced Placement or ChemCom, honors or general chemistry, you should find a large number of experiments in this book that match the scope and objectives of your course. Following each student experiment there is an extensive Teacher Information section with sample results, answers to questions, directions for preparing solutions, and other helpful hints regarding the planning and implementation of a particular experiment.

Experiments in this book can be used unchanged or they can be modified using the word-processing files provided on the accompanying CD. Students will respond differently to the design of the experiments, depending on teaching styles of their teachers, math background, previous experience using graphing calculators, and the scope and level of the chemistry course. Here are some ways to use the experiments in this book.

- Unchanged. You can photocopy the student sheets, distribute them, and have the experiments done following the procedures as they are. Many students will be more comfortable if most of the calculator steps used in data collection and analysis are included in each experiment.

- Slightly modified. The CD accompanying the book are for this purpose. Before producing student copies, you can change the directions to make them better fit your teaching circumstances. See *Appendix A* for instructions on using the CD.

- Extensively modified. This, too, can be accomplished using the accompanying CD. Some teachers will want to decrease the degree of detail in student instructions.

We hope and expect that experienced chemistry teachers will significantly modify the procedures given in this manual.

We feel it is **VERY IMPORTANT** to have your students perform Experiment 1, "Endothermic and Exothermic Reactions." This experiment has introductory details that are not included in other experiments in the book. These include specific directions for starting up the data-collection program, methods of examining data on graphs, viewing data lists, and printing graphs. The experiment can be completed quickly, leaving students plenty of time to explore other important capabilities of the probeware and data-collection program.

We also feel it is **IMPORTANT** for teachers to read the information presented in the appendices. They include valuable information that can help make you more comfortable with your initial use of this equipment. Here is a short summary of the information available in each appendix:

- *Appendix A* tells you how to use the word-processing files found on the CD.
- *Appendix B* tells you how to use TI Connect to load the EasyData App onto your calculator and capture calculator screen images.
- *Appendix C* tells you how to use Logger *Pro* 3 software to display or print graphs and data tables after importing data.
- *Appendix D* provides details about safety information.
- *Appendix E* provides information on Vernier Software & Technology products for chemistry.
- *Appendix F* provides a list of the equipment and supplies used in these experiments.

We are grateful to Pat Holmquist, Birgit Volz, and Mary Pat Randall for their patience and understanding throughout this project.

Dan D. Holmquist   Donald L. Volz   Jack Randall
dholmquist@vernier.com   jrandall@vernier.com

Endothermic and Exothermic Reactions

Many chemical reactions give off energy. Chemical reactions that release energy are called *exothermic* reactions. Some chemical reactions absorb energy and are called *endothermic* reactions. You will study one exothermic and one endothermic reaction in this experiment.

In Part I, you will study the reaction between citric acid solution and baking soda. An equation for the reaction is:

$$H_3C_6H_5O_7(aq) + 3\ NaHCO_3(s) \longrightarrow 3\ CO_2(g) + 3\ H_2O(l) + Na_3C_6H_5O_7(aq)$$

In Part II, you will study the reaction between magnesium metal and hydrochloric acid. An equation for this reaction is:

$$Mg(s) + 2\ HCl(aq) \longrightarrow H_2(g) + MgCl_2(aq)$$

OBJECTIVES

In this experiment, you will

- Study one exothermic and one endothermic reaction.
- Become familiar with using Logger *Pro.*
- Collect and display data on a graph.

Figure 1

MATERIALS

computer
Vernier computer interface
Logger*Pro*
Temperature Probe
50 mL graduated cylinder
balance

Styrofoam cup
250 mL beaker
citric acid, $H_3C_6H_5O_7$, solution
baking soda, $NaHCO_3$
hydrochloric acid, HCl, solution
magnesium, Mg

PROCEDURE

1. Obtain and wear goggles.

Part I Citric Acid plus Baking Soda

2. Place a Styrofoam cup into a 250 mL beaker as shown in Figure 1. Measure out 30 mL of citric acid solution into the Styrofoam cup. Place a Temperature Probe into the citric acid solution.

3. Connect the probe to the computer interface. Prepare the computer for data collection by opening the file "01 Endo Exothermic" from the *Chemistry with Vernier* folder of Logger*Pro*.

4. Weigh out 10.0 g of solid baking soda on a piece of weighing paper.

5. The Temperature Probe must be in the citric acid solution for at least 30 seconds before this step. Begin data collection by clicking ▶ Collect. After about 20 seconds have elapsed, add the baking soda to the citric acid solution. Gently stir the solution with the Temperature Probe to ensure good mixing. Collect data until a minimum temperature has been reached and temperature readings begin to increase. You can click on ■ Stop to end data collection or let the computer automatically end it after 300 seconds.

6. Dispose of the reaction products as directed by your teacher.

7. To analyze and print your data:

   a. Click the Statistics button, ⬚. In the statistics box that appears on the graph, several statistical values are displayed for Temp 1, including minimum and maximum. In your data table, record the maximum as the initial temperature and the minimum as the final temperature. Close the statistics box by clicking the upper-left corner of the box.

   b. To confirm the minimum and maximum temperatures, use the scroll bars in the table to scroll through the table to examine the data. Compare the minimum and maximum data points to those you recorded in the previous step.

   c. Print a copy of the table. Enter your name(s) and the number of copies.

   d. You will often want to change the scale of either axis of the graph. There are several ways to do this. To scale the temperature axis from 0 to 25°C instead of the present scaling, click the mouse on the "40" tickmark at the top of the axis. In place of the "40", type in "25" and press the ENTER key. Notice that the entire axis readjusts to the change you made. Use the same method to change the "–10" tickmark to "0". Note: A second option is to click the Autoscale button, ⬚. The computer will automatically rescale the axes for you.

   e. You can also expand any portion of the graph by zooming in on it. Select the area you want to zoom in on. Do this by moving the mouse pointer to the beginning of this section of data—press the mouse button and hold it down as you drag across the curve, leaving a rectangle. Then click the Zoom In button, ⊕. The computer will now create a new, full-size graph that includes just the region inside the rectangle. You can reverse this action by clicking the Undo Zoom button, ⊛.

f. When you again collect data in Part II of this experiment, the data will be collected as Latest run, the *most recent* set of data you have collected. The original Latest run will be lost if it is not saved or stored. Choose Store Latest Run from the Experiment menu to *store* Latest as Run 1, then save or print it later. Note that the line for Run 1 is thinner than it was for Latest. To hide the curve of your first data run, click the Temperature vertical-axis label of the graph, click More, and uncheck Run 1 Temperature. Click ⟦ OK ⟧.

Part II Hydrochloric Acid Plus Magnesium

8.   Manually rescale the vertical axis to the original temperature scale of –10 to 40°C. To do so, click the mouse on the bottom tickmark and type in "–10". Then click on the top tickmark and type in "40".

9.   Measure out 30 mL of HCl solution into the Styrofoam cup. Place the Temperature Probe into the HCl solution. Note: The Temperature Probe must be in the HCl solution for at least 45 seconds before doing Step 11. **CAUTION:** *Hydrochloric acid is caustic. Avoid spilling it on your skin or clothing. Wear chemical splash goggles at all times. Notify your teacher in the event of an accident.*

10. Obtain a piece of magnesium metal from the teacher.

11. Begin data collection by clicking ⟦▶ Collect⟧. After about 20 seconds have elapsed, add the Mg to the HCl solution. Gently stir the solution with the Temperature Probe to ensure good mixing. **CAUTION:** *Do not breathe the vapors!* Collect data until a maximum temperature has been reached and the temperature readings begin to decrease.

12. Dispose of the reaction products as directed by your teacher.

13. To analyze your Part II data:

a. Change the appearance of the graph by double-clicking anywhere on the graph bring up the Graph Options dialog. Check the box in front of Point Protector—a point protector will now outline each data point on the graph window in this trial, click the Examine button, ⟦⟧. The cursor will become a vertical line. As you move the mouse pointer across the screen, the temperature and time values corresponding to its position will be displayed in the box at the upper-left corner of the graph. Scroll across the initial 3-4 points to determine the initial temperature. Record the initial temperature in the data table. Move the mouse pointer across the peak of the temperature curve to determine the maximum temperature, and record it as the final temperature in your data table. To remove the examine box, click the upper-left corner of the box.

b. It is also possible to calculate statistics just for a portion of your collected data. To do so, you must first *select* the data you are interested in. For example, you might want to find the average (or mean) of the first few data points to use as an initial temperature, instead of using the minimum value. Select the flat portion of the curve—move the mouse pointer to time 0 and drag across the flat part of the curve. Now click the Statistics button, ⟦⟧, and note the mean temperature value in the statistics box on the graph. This value is the mean of only the selected data points. When you are done, click on the upper-left corner of the statistics box to remove it.

14. To print a graph of temperature *vs.* time showing both data runs:

a. Click the Temperature vertical-axis label of the graph. To display both temperature runs, click More, and check the Run 1 Temperature and Latest Temperature boxes. Click ⟦ OK ⟧.

b. Label both curves by choosing Text Annotation from the Insert menu, and typing "Endothermic" (or "Exothermic") in the edit box. Then drag each box to a position near its respective curve. Adjust the orientation of the arrowhead by clicking and dragging to the desired position.

c. Print a copy of the graph. Enter your name(s) and the number of copies of the graph you want.

15. Save the temperature and time data from both data runs. Choose Save As from the File menu and give the file a distinct name. As directed by your teacher, choose a location for the file, and click [Save].

DATA TABLE

	Part I	Part II
Final temperature, t_2	°C	°C
Initial temperature, t_1	°C	°C
Temperature change, Δt	°C	°C

OBSERVATIONS

PROCESSING THE DATA

1. Calculate the temperature change, Δt, for each reaction by subtracting the initial temperature, t_1, from the final temperature, t_2 ($\Delta t = t_2 - t_1$).

2. Tell which reaction is exothermic. Explain.

3. Which reaction had a negative Δt value? Is the reaction endothermic or exothermic? Explain.

4. For each reaction, describe three ways you could tell a chemical reaction was taking place.

5. Which reaction took place at a greater rate? Explain your answer.

TEACHER INFORMATION

Endothermic and Exothermic Reactions

1. The student pages with complete instructions for data collection using LabQuest App, Logger *Pro* (computers), EasyData or DataMate (calculators), and DataPro (Palm handhelds) can be found on the CD that accompanies this book. See *Appendix A* for more information.

2. This experiment serves as an introduction to the use of the data-collection software and temperature probes. In the procedure, students are encouraged to explore many of the menu options available in the software.

3. Instruct your students to weigh out the solid baking soda on a piece of weighing paper.

4. The solutions can be prepared as follows:

   1.5 M citric acid (288.2 g $H_3C_6H_5O_7$ per 1 L solution) Hazard Code: D—Relatively non-hazardous. Alternatively, 315.2 g $H_3C_6H_5O_7 \cdot H_2O$ per 1 L solution. **HAZARD ALERT:** Severe eye irritant. Hazard Code: D—Relatively non-hazardous.

   1.0 M HCl (85.6 mL concentrated reagent per 1 L solution) **HAZARD ALERT:** Highly toxic by ingestion or inhalation; severely corrosive to skin and eyes. Hazard Code: A—Extremely hazardous.

5. Provide pre-cut strips of magnesium ribbon weighing about 0.10 g. **HAZARD ALERT:** Flammable solid; burns with an intense flame; keep either dry sand or Flinn *Met-L-X®* available to use as a fire extinguisher. Hazard Code: C—Somewhat hazardous.

   The hazard information reference is: Flinn Scientific, Inc., *Chemical & Biological Catalog Reference Manual*, (800) 452-1261, www.flinnsci.com. See *Appendix D* of this book, *Chemistry with Vernier*, for more information.

6. We recommend placing the Styrofoam cup in a 250 mL beaker because it adds stability to the cup and provides extra insulation.

7. If you are using the Stainless Steel Temperature Probe, Go!Temp or the EasyTemp, caution your students not to leave the probe in the solutions for extended periods of time after Part II is completed; mild discoloration of the probe will occur after 20–25 minutes in 1 M HCl.

8. If you are using calculators and EasyData for data collection, remind students to collect data for the entire 5 minutes. If they stop data collection early, they will not be able to view both graphs simultaneously.

ANSWERS TO QUESTIONS

2. The Part II reaction was exothermic. The temperature went up as energy was released by the reaction.

3. The Part I reaction had a negative Δt (–19.6°C). This means there was a decrease in temperature, which was due to energy being absorbed by the endothermic reaction.

4. Some examples of evidence of a chemical reaction in Part I are:

   - The temperature decreased.
   - The solid baking soda disappeared.
   - A gas was produced.

   Examples of Part II reaction evidence are:

   - The temperature increased.
   - The magnesium metal disappeared.
   - A gas was produced.

5. Reaction II took place at a greater rate. It reached a maximum temperature after 80 seconds and Reaction I reached a minimum after 205 seconds. Both reactions were near completion at these times (bubbling had nearly ceased).

SAMPLE RESULTS

	Part I	Part II
Final temperature, t_2	1.2°C	35.0°C
Initial temperature, t_1	20.8°C	21.1°C
Temperature change, Δt	–19.6°C	13.9°C

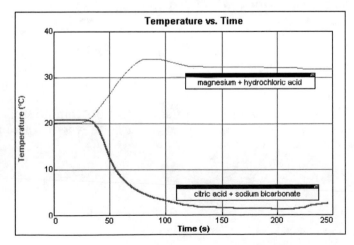

Endothermic and Exothermic Reactions

Freezing and Melting of Water

Freezing temperature, the temperature at which a substance turns from liquid to solid, and melting temperature, the temperature at which a substance turns from a solid to a liquid, are characteristic physical properties. In this experiment, the cooling and warming behavior of a familiar substance, water, will be investigated. By examining graphs of the data, the freezing and melting temperatures of water will be determined and compared.

OBJECTIVES

In this experiment, you will

- Collect temperature data during the freezing and melting of water.
- Analyze graphs to determine the freezing and melting temperatures of water.
- Determine the relationship between the freezing and melting temperatures of water.

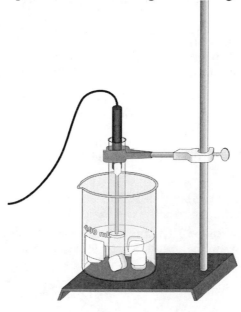

Figure 1

MATERIALS

computer	400 mL beaker
Vernier computer interface	water
Logger*Pro*	10 mL graduated cylinder
Temperature Probe	ice
ring stand	salt
utility clamp	stirring rod
test tube	

PROCEDURE

Part I: Freezing

1. Fill a 400 mL beaker 1/3 full with ice, then add 100 mL of water.

2. Put 5 mL of water into a test tube and use a utility clamp to fasten the test tube to a ring stand. The test tube should be clamped above the water bath. Place a Temperature Probe into the water inside the test tube.

3. Connect the probe to the computer interface. Prepare the computer for data collection by opening the file "02 Freeze Melt Water" from the *Chemistry with Vernier* folder of Logger *Pro*.

4. When everything is ready, click ▶ Collect to begin data collection. Then lower the test tube into the ice-water bath.

5. Soon after lowering the test tube, add 5 spoons of salt to the beaker and stir with a stirring rod. Continue to stir the ice-water bath during Part I. **Important:** Stir enough to dissolve the salt.

6. Slightly, but continuously, move the probe during the first 10 minutes of Part I. Be careful to keep the probe in, and not above, the ice as it forms. When 10 minutes have gone by, stop moving the probe and allow it to freeze into the ice. Add more ice cubes to the beaker as the original ice cubes get smaller.

7. When 15 minutes have passed, data collection will stop. Keep the test tube *submerged* in the ice-water bath until Step 10.

8. On the displayed graph, analyze the flat part of the curve to determine the freezing temperature of water:

   a. Move the mouse pointer to the beginning of the graph's flat part. Press the mouse button and hold it down as you drag across the flat part to *select* it.
   b. Click on the Statistics button, 📊. The mean temperature value for the selected data is listed in the statistics box on the graph. Record this value as the freezing temperature in your data table.
   c. To remove the statistics box, click on the upper-left corner of the box.

Part II: Melting

9. Prepare the computer for data collection. From the Experiment menu, choose Store Latest Run. This stores the data so it can be used later. To hide the curve of your first data run, click the Temperature vertical-axis label of the graph, click More, and uncheck the Run 1 Temperature box. Click OK.

10. Click ▶ Collect to begin data collection. Then raise the test tube and fasten it in a position above the ice-water bath. Do not move the Temperature Probe during Part II.

11. Dispose of the ice water as directed by your teacher. Obtain 250 mL of warm tap water in the beaker. When 12 minutes have passed, lower the test tube and its contents into this warm-water bath.

12. When 15 minutes have passed, data collection will stop.

13. On the displayed graph, analyze the flat part of the curve to determine the melting temperature of water:

    a. Move the mouse pointer to the beginning of the graph's flat part. Press the mouse button and hold it down as you drag across the flat part to *select* it.

    b. Click the Statistics button, ⬚. The mean temperature value for the selected data is listed in the statistics box on the graph. Record this value as the melting temperature in your data table.

    c. To remove the statistics box, click on the upper-left corner of the box.

14. To print a graph of temperature *vs.* time showing both data runs:

    a. Click the Temperature vertical-axis label of the graph. To display both temperature runs, click More, and check the Run 1 Temperature and Latest Temperature boxes. Click   OK .

    b. Label both curves by choosing Text Annotation from the Insert menu, and typing "Freezing Curve" (or "Melting Curve") in the edit box. Then drag each box to a position near its respective curve. Adjust the orientation of the arrowhead by clicking and dragging to the desired position.

    c. Print a copy of the graph. Enter your name(s) and the number of copies of the graph you want.

OBSERVATIONS

DATA TABLE

Freezing temperature of water	°C
Melting temperature of water	°C

PROCESSING THE DATA

1. What happened to the water temperature during freezing? During melting?

2. According to your data and graph, what is the freezing temperature of water? The melting temperature? Express your answers to the nearest 0.1°C.

3. How does the freezing temperature of water compare to its melting temperature?

4. Tell if the *kinetic energy* of the water in the test tube increases, decreases, or remains the same in each of these time segments during the experiment.

    a. when the temperature is changing at the beginning and end of Part I
    b. when the temperature remains constant in Part I
    c. when the temperature is changing at the beginning and end of Part II
    d. when the temperature remains constant in Part II

5. In those parts of Question 4 in which there was no kinetic energy change, tell if *potential energy* increased or decreased.

Experiment

2

Freezing and Melting of Water

1. The student pages with complete instructions for data-collection using LabQuest App, Logger *Pro* (computers), EasyData or DataMate (calculators), and DataPro (Palm handhelds) can be found on the CD that accompanies this book. See *Appendix A* for more information.

2. This entire experiment requires a full 45–50 minute period. Students should have done Experiment 1 before this one. Be sure to go over this experiment well with your students, especially if it is one of the first computer-interfaced experiments they have done. As the Sample Results below show, this procedure can give excellent results.

3. The stored calibration for all Vernier Temperature Probes works well for this experiment—the freezing and melting temperatures of water should be within ± 0.2°C of 0°C using these calibrations.

4. Test tubes size 20 × 150 mm work well. Sizes 25 × 150 mm and 18 × 150 mm work, too.

5. A water sample size of 5 mL works well. Larger samples will take more time than is recommended in this procedure.

6. As shown in the first graph in the Sample Results, many of the samples will supercool. Stirring will bring the super-cooled water to the melting temperature plateau.

7. If you are using calculators and EasyData for data collection, remind students to collect data for the entire 15 minutes. If they stop data collection early, they will not be able to view both graphs simultaneously.

ANSWERS TO QUESTIONS

1. The water temperature stayed constant near 0°C during freezing and melting.

2. The expected value is 0°C for both the freezing and melting temperatures, but answers will vary slightly.

3. The freezing and melting temperatures of water are the same.

4. a. Average kinetic energy decreases with the temperature decrease at the beginning and end of Part I.
   b. Since there is no temperature change during freezing, average kinetic energy remains constant.
   c. Average kinetic energy increases with the temperature increase at the beginning and end of Part II.
   d. Since there is no temperature change during melting, average kinetic energy is constant.

5. b. Potential energy decreased during freezing
   d. Potential energy increased during melting.

SAMPLE DATA

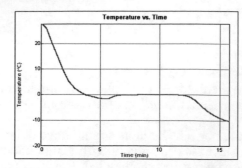

Part I: Freezing Water

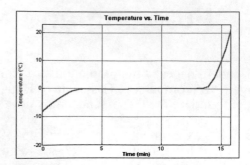

Part II: Melting Water

Another Look
at Freezing Temperature

In the experiment, "Freezing and Melting of Water," you saw that the temperature of *pure* water remained constant at its freezing temperature as it froze and melted. Using a computer-interfaced Temperature Probe, you will now observe what happens when a different pure substance, phenyl salicylate, freezes. Using a second Temperature Probe and sample, you will also see the effect on freezing temperature when a small amount of another substance, benzoic acid, is dissolved in the phenyl salicylate.

OBJECTIVES

In this experiment, you will

- Observe what happens when phenyl salicylate freezes.
- See the effect on the freezing temperature when a small amount of benzoic acid is dissolved in the phenyl salicylate.

MATERIALS

computer	ring stand
Vernier computer interface	2 utility clamps
Logger*Pro*	test tube with phenyl salicylate
2 Temperature Probes	test tube with phenyl salicylate & benzoic acid
1 liter beaker	2 crystals of solid phenyl salicylate
thermometer (optional)	

PROCEDURE

1. Obtain and wear goggles.

2. Connect the probes to the computer interface. Prepare the computer for data collection by opening the file "03 Another Look Freezing" from the *Chemistry with Vernier* folder of Logger*Pro*.

3. Fill a 1 liter beaker about 4/5 full with water at a temperature in the range 30–32°C. This temperature can be obtained by blending hot and cold tap water. This water temperature should be measured with a thermometer or Temperature Probe. Place the 1 liter beaker on the base of the ring stand.

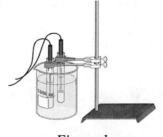

Figure 1

4. Obtain 2 crystals of solid phenyl salicylate to use in Step 9.

5. Identify the Channel 1 and Channel 2 Temperature Probes. The Channel 1 Temperature Probe will be used with pure phenyl salicylate, and the Channel 2 probe with the benzoic acid-phenyl salicylate mixture. **Note:** Plan ahead for quick positioning of the probes and test tubes in Step 7!

6. When everything is ready, use two utility clamps to obtain test tubes containing the two hot liquids from your teacher. While one team member obtains the melted phenyl salicylate, another can get the melted benzoic acid-phenyl salicylate mixture. Fasten the utility clamps at the top of the test tubes to carry them back to your lab station. **CAUTION:** *Be careful not to spill the hot liquids and do not touch the bottom of the test tubes.*

7. *Immediately* insert the Channel 1 Temperature Probe into the hot phenyl salicylate and the Channel 2 probe into the hot benzoic acid-phenyl salicylate mixture. Then attach the clamps to the ring stand with the test tubes still above the water. About 30-45 seconds are required for the probes to warm up to the temperature of their surroundings and give correct temperature readings. During this time, watch the live temperature values displayed in the meter. When the temperatures start to drop, click ▶ Collect to begin data collection. Lower one test tube, then the other, into the water bath as shown in Figure 1.

8. Make sure the water level outside the test tubes is higher than the liquid levels inside the test tubes. The Channel 1 probe is now monitoring the phenyl salicylate as Temperature 1, and the Channel 2 probe is monitoring the benzoic acid-phenyl salicylate mixture as Temperature 2. If the temperature at the start of the graph was not above 50°C for either one of your Temperature Probes, obtain two more samples and begin again.

9. With a very slight up and down motion of both probes, continuously stir the liquids during the cooling. Hold the tops of the probes and *not* their wires! If no solid appears in either test tube by the time the temperature drops below 39°C, add a crystal of solid phenyl salicylate to the liquid(s) with no solid.

10. When either probe begins to stick in the phenyl salicylate, stop stirring. Continue on with the experiment until both temperatures have dropped below 31°C or until 20 minutes have passed. Click ■ Stop to end data collection or wait for it to end automatically after 20 minutes.

11. When you have finished collecting data, if a Temperature Probe has frozen into the phenyl salicylate, use the hot water bath provided by your teacher to melt it out. Do *not* attempt to pull the probe out—this might damage it!

12. Return the test tubes containing the phenyl salicylate and the benzoic acid-phenyl salicylate mixture to the places directed by your teacher.

13. To print a graph of temperature *vs.* time for the two cooling curves:

    a. Label each curve by choosing Text Annotation from the Insert menu, and typing "Phenyl salicylate" (or "Benzoic acid-phenyl salicylate") in the edit box. Then drag each box to a position near its respective curve. Adjust the orientation of the arrowhead by clicking and dragging it to the desired position.
    b. Print a copy of the graph. Enter your name(s) and the number of copies of the graph you want.

14. Proceed directly to Step 1 of Processing the Data.

PROCESSING THE DATA

1. Find the freezing temperature of pure phenyl salicylate:

   a. To hide the curve of the Channel 2 probe, click the Temperature vertical-axis label of the graph, and choose Temperature 1.

   b. Move the mouse pointer to the beginning of the graph's flat part. Press the mouse button and hold it down as you drag across the flat section to *select* it.

   c. Click on the Statistics button, 🔲. The mean temperature value for the selected data is listed in the statistics box on the graph. Record this value as the freezing temperature of pure phenyl salicylate in your data table.

   d. To remove the statistics box, click on the upper-left corner of the box.

2. Find the freezing temperature of the benzoic acid-phenyl salicylate mixture:

   a. Click the Temperature vertical-axis label of the graph and choose Temperature 2.

   b. To find the freezing point of a mixture, you need to determine the temperature at which it first started to freeze. Unlike pure phenyl salicylate, your data may show a linear, but gradually decreasing change in temperature during the time period when freezing takes place. Move the mouse pointer to the beginning of the graph's sloping linear sloping section. Press the mouse button and hold it down as you drag only the sloping linear region of the curve.

   c. Click on the Linear Fit button, 🔲. A regression line will fit the section of data you selected.

   d. Choose Interpolate from the Analyze menu. A vertical cursor now appears on the graph. As you move the mouse pointer along the regression line, the cursor's x and y coordinates are displayed in the floating box (x is time and y is temperature). Move the cursor left along the regression line to the point where it intersects the initial steep temperature drop of the cooling curve—this is the temperature when freezing began. Record the initial freezing temperature of the mixture in your data table.

3. How are the curves on the two graphs different? You should see at least two effects that dissolving benzoic acid in phenyl salicylate had on the freezing process.

4. Based on this lab, what method could a chemist use to determine whether an unknown liquid is pure or has an impurity dissolved in it?

DATA TABLE

Freezing temperature, phenyl salicylate	°C
Freezing temperature, benzoic acid-phenyl salicylate mixture	°C

TEACHER INFORMATION

Another Look at Freezing Temperature

1. The student pages with complete instructions for data-collection using LabQuest App, Logger *Pro* (computers), EasyData or DataMate (calculators), and DataPro (Palm handhelds) can be found on the CD that accompanies this book. See *Appendix A* for more information.

2. This activity is ideally performed with a LabQuest, LabPro, or CBL 2 and two Stainless Steel Temperature Probes. It is possible to perform this activity with one Temperature Probe, but it will take additional time.

   If you use only one Temperature Probe, have the students do one trial using phenyl salicylate, followed by a second trial using a benzoic acid-phenyl salicylate mixture. In order to complete the experiment in one class period, you may have to reduce the length of the trial to 12–15 minutes.

   Note: Do not use the white TI Temperature Probe that was shipped with the original CBL; using it in phenyl salicylate or benzoic acid-phenyl salicylate can permanently damage the probe.

3. Phenyl salicylate is also known as salol and its freezing temperature is about 41°C.

4. The stored calibration for the Stainless Steel Temperature Probe or Direct-Connect Temperature Probe works well for this experiment.

5. Use 20 × 150 or 25 × 150 mm test tubes.

6. Add 15 g of phenyl salicylate to each test tube. Add about 0.50 g of solid benzoic acid to 15 g of phenyl salicylate for the other sample. Two hot-water baths on separate hot plates adjusted to a low setting (one that maintains the water bath at a temperature of 80–90°C) can be used to melt the solids and maintain them in the liquid state. Label test tubes to ensure that they are returned to the proper water bath.

7. In Processing the Data students use two methods to analyze the data points on the graph to determine the freezing temperature. Use graphs similar to Figures 1 and 2 to show students how to find freezing temperatures of the pure liquid and the solution. We purposely do not include these graphs in the experiment so students can discover the shapes of the cooling curves themselves. Students should be able to quickly determine the freezing temperature of pure phenyl salicylate by examining the data points along the freezing plateau (See Figure 1). The benzoic acid-phyenyl salicylate mixture is more difficult, especially if supercooling occurs. Move the cursor to the approximate point where the linear sloping plateau would intersect the steep slope of the cooling liquid, as if supercooling had not occurred (see Figure 2).

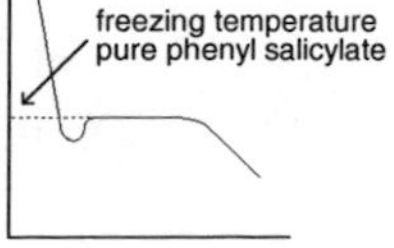

Figure 1

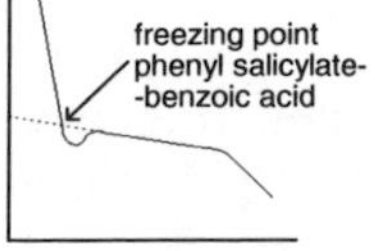

Figure 2

If you feel your students are not ready for the data analysis described in Steps 1 and 2, an easier method is to simply have them examine the graph and move the cursor along the curve to the temperature points indicated in Figure 1. The melting temperature can be taken as the point where the extrapolated dotted line intersects the steep initial temperature slope.

8. To find the freezing temperature of pure phenyl salicylate, shown in Figure 3, students will follow this procedure (Step 1 of Processing the Data):

   a. Select the region of data for the flat section of the graph (see Figure 3).

   b. Use the Statistics analysis tool of your program, and determine the mean temperature value for the selected data. Record this value as the freezing temperature of pure phenyl salicylate in your data table.

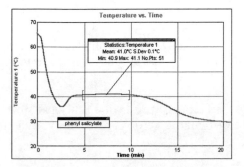

Figure 3

9. To find the freezing temperature of the benzoic acid-phenyl salicylate mixture, shown in Figure 4, students will follow this procedure (Step 2 of Processing the Data):

   a. Select the region of data for just the sloping linear region of the curve (see Figure 4).

   b. Use the Linear Fit tool of your data-collection program to fit the selected data with a regression line.

   c. Use the Interpolate tool of your data-collection program to determine the point where the linear fit intersects the initial steep temperature drop of the cooling curve—this is the temperature when freezing began. Record the initial freezing temperature of the mixture in your data table (see the interpolated value of 39.7°C in Figure 4).

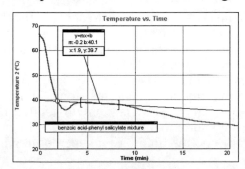

Figure 4

10. The starting temperature of the water surrounding the test tubes should have a reading of 50°C or greater. This ensures a portion of the graph showing the drop down to the 41°C freezing temperature of phenyl salicylate.

11. Students may need to use a seed crystal if no solid appears in either test tube by the time the temperature drops below 39°C. Many of the samples will supercool. It is best to seed soon after supercooling begins. Provide students with seed crystals in advance.

12. Remelting phenyl salicylate in a test tube that had no probe or thermometer present in it during freezing often results in a broken test tube. It is a good idea to keep the melted phenyl salicylate in a warm water bath for reuse in subsequent periods. After the last class completes the experiment, you can pour the melted phenyl salicylate or benzoic acid-phenyl salicylate mixture solution into two sturdy plastic bags. When it is solidified, the phenyl salicylate can be crushed for reuse.

13. **HAZARD ALERTS:**

Benzoic acid: Moderately toxic by ingestion; irritates eyes, skin and respiratory tract; combustible. Hazard Code: C—Somewhat hazardous.

Phenyl salicylate: Moderately toxic by ingestion. Hazard Code: C—Somewhat hazardous.

The hazard information reference is: Flinn Scientific, Inc., *Chemical & Biological Catalog Reference Manual,* (800) 452-1261, www.flinnsci.com. See *Appendix D* of this book, *Chemistry with Vernier*, for more information.

SAMPLE RESULTS

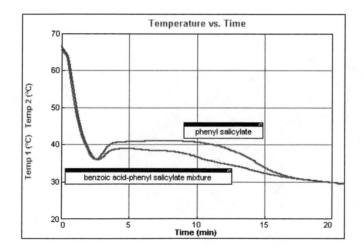

ANSWERS TO QUESTIONS

1. 41.0°C

2. 39.7°C

3. The freezing temperature of the benzoic acid-phenyl salicylate mixture is lower than the freezing temperature of the pure phenyl salicylate. There is almost no change in temperature during the freezing of pure phenyl salicylate. During the freezing of the solution containing benzoic acid, there was a steady drop in temperature.

4. Cool the unknown liquid until it starts to freeze. If the temperature remains constant during the time the liquid is freezing, it must be pure. If the temperature slowly drops during the freezing process, the liquid must have an impurity dissolved in it.

Heat of Fusion for Ice

Melting and freezing behavior are among the characteristic properties that give a pure substance its unique identity. As energy is added, pure solid water (ice) at 0°C changes to liquid water at 0°C.

In this experiment, you will determine the energy (in joules) required to melt one gram of ice. You will then determine the molar heat of fusion for ice (in kJ/mol). Excess ice will be added to warm water, at a known temperature, in a Styrofoam cup. The warm water will be cooled down to a temperature near 0°C by the ice. The energy required to melt the ice is removed from the warm water as it cools.

To calculate the heat that flows from the water, you can use the relationship

$$q = C_p \cdot m \cdot \Delta t$$

where q stands for heat flow, C_p is specific heat, m is mass in grams, and Δt is the change in temperature. For water, C_p is 4.18 J/g°C.

OBJECTIVES

In this experiment, you will

- Determine the energy (in Joules) required to melt one gram of ice.
- Determine the molar heat of fusion for ice (in kJ/mol).

MATERIALS

computer	Styrofoam cup
Vernier computer interface	ring stand
Logger Pro	utility clamp
Temperature Probe	ice cubes
250 mL beaker	stirring rod
100 mL graduated cylinder	warm water
tongs	

PROCEDURE

1. Connect the probe to the computer interface. Prepare the computer for data collection by opening the file "04 Heat of Fusion" from the *Chemistry with Vernier* folder.

2. Place a Styrofoam cup into a 250 mL beaker as shown in Figure 1.

3. Use a utility clamp to suspend the Temperature Probe on a ring stand as shown in Figure 1.

4. Use a 100 mL graduated cylinder to obtain 100.0 mL of water at about 60°C from your instructor. Record this as V_1.

5. Obtain 7 or 8 large ice cubes.

6. Lower the Temperature Probe into the warm water (to about 1 cm from the bottom).

7. Click ▶Collect to begin data collection. Wait until the temperature reaches a maximum (it will take a few seconds for the cold probe to reach the temperature of the warm water). This maximum will determine the initial temperature, t_1, of the water. As soon as this maximum temperature is reached, fill the Styrofoam cup with ice cubes. Shake excess water from the ice cubes before adding them (or dry with a paper towel). Record the maximum temperature, t_1, in your data table.

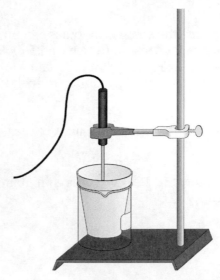

Figure 1

8. Use a stirring rod to stir the mixture as the temperature approaches 0°C. **Important:** As the ice melts, add more large ice cubes to keep the mixture full of ice!

9. When the temperature reaches about 4°C, quickly remove the unmelted ice (using tongs). Continue stirring until the temperature reaches a minimum (and begins to rise). This minimum temperature is the final temperature, t_2, of the water. Record t_2 in your data table. Click ■Stop when you have finished collecting data.

10. Use the 100 mL graduated cylinder to measure the volume of water remaining in the Styrofoam cup to the nearest 0.1 mL. Record this as V_2.

11. You can confirm your data by clicking the Statistics button, ⌷. The minimum temperature (t_2) and maximum temperature (t_1) are listed in the floating box on the graph.

PROCESSING THE DATA

1. Use the equation $\Delta t = t_2 - t_1$ to determine Δt, the change in water temperature.

2. Subtract to determine the volume of ice that was melted ($V_2 - V_1$).

3. Find the mass of ice melted using the volume of melt (use 1.00 g/mL as the density of water).

4. Use the equation given in the introduction of this experiment to calculate the energy (in joules) released by the 100 g of liquid water as it cooled through Δt.

5. Now use the results obtained above to determine the heat of fusion—the energy required to melt one gram of ice (in J/g H_2O).

6. Use your answer to Step 5 and the molar mass of water to calculate the molar heat of fusion for ice (in kJ/mol H_2O).

7. Find the percent error for the molar heat of fusion value in Step 6. The accepted value for molar heat of fusion is 6.01 kJ/mol.

DATA AND CALCULATIONS

Initial water temperature, t_1		°C
Final water temperature, t_2		°C
Change in water temperature, Δt		°C
Final water volume, V_2		mL
Initial water volume, V_1		mL
Volume of melt		mL

Mass of ice melted	
	g
Heat released by cooling water ($q = C_p \cdot m \cdot \Delta t$)	
	J
J/g ice melted (heat of fusion)	
	J/g
kJ/mol ice melted (molar heat of fusion)	
	kJ/mol
Percent error	
	%

Heat of Fusion for Ice

1. The student pages with complete instructions for data-collection using LabQuest App, Logger *Pro* (computers), EasyData or DataMate (calculators), and DataPro (Palm handhelds) can be found on the CD that accompanies this book. See *Appendix A* for more information.

2. Have hot water at approximately 70°C available on a hot plate. Such a temperature can be maintained at a low setting.

3. By the time students get the water measured and ready for data collection, its temperature will have dropped to 50-55°C. Using *plenty* of *large* ice cubes works best. This speeds the process and minimizes error.

4. A beginning temperature of 50°C is generally recommended to provide equal temperature ranges above and below room temperature.

5. The ice cubes should be wet; that is, at melting temperature. They therefore need to be removed from the freezer 15–20 minutes before they will be used.

6. When students are positioning a Stainless Steel Temperature Probe for this experiment, they can simply tighten the utility clamp around the probe handle.

   If they are using an earlier Direct-Connect Temperature Probe, however, a 1-hole rubber stopper (No. 2 to 6) should be used to fasten the probe in a utility clamp. Sliding the probe through the hole might damage the probe. Instead, cut the stopper open with a single cut from the center hole to the outside. The slit stopper will then readily wrap around the top of the probe.

7. We recommend placing the Styrofoam cup in a 250 mL beaker because it adds stability to the cup and provides extra insulation.

SAMPLE RESULTS

Initial water temperature, t_1	53.0°C
Final water temperature, t_2	1.9°C
Change in water temperature, Δt	51.1°C
Final water volume, V_2	164.5 mL
Initial water volume, V_1	100.0 mL
Volume of melt	64.5 mL

Mass of ice melted

$$(64.5 \text{ mL})(1.00 \text{ g/mL}) =$$

64.5 g

Heat released by cooling water ($q = C_p \cdot m \cdot \Delta t$)

$$(4.18 \text{ J/g°C})(100 \text{ g})(51.1°C) =$$

21400 J

J/g ice melted (heat of fusion)

$$\frac{21400 \text{ J}}{64.5 \text{ g}} =$$

332 J/g

kJ/mol ice melted (molar heat of fusion)

$$(332 \text{ J/g})(18.0 \text{ g/mol}) = 5{,}980 \text{ J/mol} =$$

5.98 kJ/mol

Percent error

$$\frac{|6.01-5.98|}{|6.01|} \times 100 =$$

0.50 %

Find the Relationship:
An Exercise in Graphing Analysis

In several laboratory investigations you do this year, a primary purpose will be to find the mathematical relationship between two variables. For example, you might want to know the relationship between the pressure exerted by a gas and its temperature. In one experiment you do, you will be asked to determine the relationship between the volume of a confined gas and the pressure it exerts. A very important method for determining mathematical relationships in laboratory science makes use of graphical methods.

OBJECTIVE

In this experiment, you will determine several mathematical relationships using graphical methods.

EXAMPLE 1

Suppose you have these four ordered pairs, and you want to determine the relationship between x and y:

x	y
2	6
3	9
5	15
9	27

The first logical step is to make a graph of *y* versus *x*.

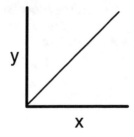

Since the shape of the plot is a straight line that passes through the origin (0,0), it is a simple *direct* relationship. An equation is written showing this relationship: $y = k \cdot x$. This is done by writing the variable from the vertical axis (dependent variable) on the left side of the equation, and then equating it to a proportionality constant, k, multiplied by x, the independent variable. The constant, k, can be determined either by finding the slope of the graph or by solving your equation for k ($k = y/x$), and finding k for one of your ordered pairs. In this simple example, $k = 6/2 = 3$. If it is the correct proportionality constant, then you should get the same k value by dividing any of the y values by the corresponding x value. The equation can now be written:

$$y = 3 \cdot x \quad (y \text{ varies directly with } x)$$

EXAMPLE 2

Consider these ordered pairs:

x	y
1	2
2	8
3	18
4	32

First plot y versus x. The graph looks like this:

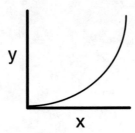

Since this graph is not a straight line passing through the origin, you must make another graph. It appears that y increases as x increases. However, the increase is not proportional (direct). Rather, y varies *exponentially* with x. Thus y might vary with the *square* of x or the *cube* of x. The next logical plot would be y versus x^2. The graph looks like this:

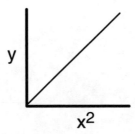

Since this plot is a straight line passing through the origin, y varies with the square of x, and the equation is:

$$y = k \cdot x^2$$

Again, place y on one side of the equation and x^2 on the other, multiplying x^2 by the proportionality constant, k. Determine k by dividing y by x^2:

$$k = y/x^2 = 8/(2)^2 = 8/4 = 2$$

This value will be the same for any of the four ordered pairs, and yields the equation:

$$y = 2 \cdot x^2 \quad (y \text{ varies directly with the square of } x)$$

EXAMPLE 3

x	y
2	24
3	16
4	12
8	6
12	4

A plot of y versus x gives a graph that looks like this:

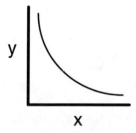

A graph with this curve always suggests an inverse relationship. To confirm an inverse relationship, plot the reciprocal of one variable versus the other variable. In this case, y is plotted versus the reciprocal of x, or $1/x$. The graph looks like this:

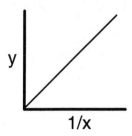

Since this graph yields a straight line that passes through the origin (0,0), the relationship between x and y is inverse. Using the same method we used in examples 1 and 2, the equation would be:

$$y = k(1/x) \text{ or } y = k/x$$

To find the constant, solve for k ($k = y \cdot x$). Using any of the ordered pairs, determine k:

$$k = 2 \times 24 = 48$$

Thus the equation would be:

$$y = 48/x \quad (y \text{ varies inversely with } x)$$

EXAMPLE 4

The fourth and final example has the following ordered pairs:

x	y
1.0	48.00
1.5	14.20
2.0	6.00
3.0	1.78
4.0	0.75

A plot of *y* versus *x* looks like this:

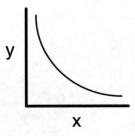

Thus the relationship must be inverse. Now plot *y* versus the reciprocal of *x*. The plot of *y* versus 1/*x* looks like this:

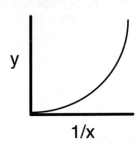

Since this graph is not a straight line, the relationship is not just inverse, but rather inverse square or inverse cube.

The next logical step is to plot *y* versus $1/x^2$ (inverse square). The plot of this graph is shown below. The line still is not straight, so the relationship is not inverse square.

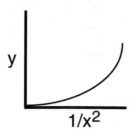

Finally, try a plot of y versus $1/x^3$. Aha! This plot comes out to be a straight line passing through the origin.

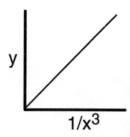

This must be the correct relationship. The equation for the relationship is:

$$y = k(1/x^3) \ \text{ or } \ y = k/x^3$$

Now, determine a value for the constant, k. For example, $k = y \cdot x^3 = (6)(2)^3 = 48$. Check to see if it is constant for other ordered pairs. The equation for this relationship is:

$$y = 48/x^3 \ (y \text{ varies inversely with the cube of } x)$$

MATERIALS

computer
Logger *Pro*

PROCEDURE

1. Obtain three problems from your instructor to solve using the graphical method described in the introduction. Follow the procedure in Steps 2-9 to find the mathematical relationship for the data pairs in each problem.

2. Begin by opening the file "05 Find the Relationship" from the *Chemistry with Vernier* folder of Logger *Pro*.

3. Enter the data pairs in the table.

    a. Click on the first cell in the x column in the table. Type in the x value for the first data pair.
    b. Click on the first cell in the y column. Type in the y value for the first data pair.
    c. Continue in this manner to enter the remaining data pairs.

4. Examine the shape of the curve in the graph. If the graph is curved (varies inversely or exponentially), proceed as described in the introduction of this experiment. To do this using Logger*Pro*, it is necessary to create a new column of data, $x^{\wedge}n$, where x represents the original x column in the Table window, and n is the value of the exponent:

   a. Choose New Calculated Column from the Data menu.

   b. Enter a Name that corresponds to the formula you will enter (e.g., "$x^{\wedge}2$", "$x^{\wedge}-1$"). Use an exponent of "2" or "3" for a power that increases exponentially, "–1" for the reciprocal of n, "–2" for inverse square, or "–3" for inverse cube. Leave the Short Name and Unit boxes empty.

   c. Enter the correct formula for the column, $(x^{\wedge}n)$ in the Equation edit box. To do this, select "x" from the Variables list. Following "x" in the Equation edit box, type in "$^{\wedge}$", then type in the value for the exponent, n, that you used in the previous step. Click [**Done**]. According to the exponent of n you entered, a corresponding set of calculated values will appear in a modified column in the table.

   d. Click on the horizontal-axis label, select "$x^{\wedge}n$". You should now see a graph of y *vs.* $x^{\wedge}n$. To autoscale both axes starting with zero, double-click in the center of the graph to view Graph Options, click the Axis Options tab, and select Autoscale from 0 from the scaling menu for both axes. Click [**Done**].

5. To see if you made the correct choice of exponents:

   a. If a straight line results, you have made the correct choice—proceed to Step 6. If it is still curved, double-click on the calculated column, $x^{\wedge}n$, heading. Decide on a new value for n, then edit the value of n that you originally entered (in the $x^{\wedge}n$ formula in the Equation edit box). Change the exponent in the Name. Click [**Done**].

   b. A new set of values for this power of x will appear in the modified column; these values will automatically be plotted on the graph.

   c. You should now see a graph of y *vs.* $x^{\wedge}n$. If necessary, autoscale both axes starting from zero.

   d. If the points are in a straight line, proceed to Step 6. If not, repeat the Step-5 procedure until a straight line is obtained.

6. After you have obtained a straight line, click the Linear Fit button, ⌐. The regression line is calculated by the computer as a best-fit straight line passing through or near the data points, and will be shown on the graph.

7. Since you will need to use the original data pairs in Processing the Data, record the x and y values, the value of n used in your final graph, and the problem number in the Data and Calculations table (or, if directed by your instructor, print a copy of the table).

8. To print your linear graph of y *vs.* $x^{\wedge}n$:

   a. Label the curves by choosing Text Annotation from the Insert menu, and typing the number of the problem you just solved in the edit box (e.g., *Problem 23*). Drag the box to a position near the curve. Adjust the position of the arrowhead by clicking and dragging it.

   b. Print a copy of the graph. Enter your name(s) and the number of copies of the graph.

9. To confirm that you made the right choice for the exponent, n, you can use a second method. Instead of a *linear* regression plot of y *vs.* x^n, you can create a *power* regression curve on the original plot of y *vs.* x. Using the method described below, you can also calculate a value for a and n in the equation, $y = A \cdot x^{\wedge}n$.

   a. To return to the original plot of y *vs.* x, click on the horizontal-axis label, and select "x". Remove the linear regression and annotation floating boxes.

b. To return to the original plot of y *vs.* x, click on the horizontal-axis label, and select "x". Remove the linear regression and annotation floating boxes.

c. Click the Curve Fit button, $\boxed{\approx}$.

d. Choose your mathematical relationship from the list at the lower left: Use Variable Power (y = Ax^n). To confirm that the exponent, *n*, is the same as the value you recorded earlier, enter the value of *n* in the Power edit box at the bottom. Click $\boxed{\text{Try Fit}}$.

e. A best-fit curve will be displayed on the graph. The curve should match up well with the points, if you made the correct choice. If the curve does not match up well, try a different power and click $\boxed{\text{Try Fit}}$ again. When the curve has a good fit with the data points, then click $\boxed{\text{OK}}$.

f. (Optional) Print a copy of the graph, with the curve fit still displayed. Enter your name(s) and the number of copies of the graph you want, then click $\boxed{\text{OK}}$.

10. To do another problem, reopen "05 Find the Relationship." **Important:** Click on the No button when asked if you want to save the changes to the previous problem. Repeat Steps 3-9 for the new problem.

PROCESSING THE DATA

1. Using *x*, *y*, and *k*, write an equation that represents the relationship between y and x for each problem. Use the value for *n* that you determined from the graphing exercise. Write your final answer using only positive exponents. For example, if $y = k{\cdot}x^{-2}$, then rewrite the answer as: $y = k/x^2$. See the Data and Calculations table for examples.

2. Solve each equation for *k*. Then calculate the numerical value of *k*. Do this for at least two ordered pairs, as shown in the example, to confirm that k is really constant. See the Data and Calculations table for examples.

3. Rewrite the equation, using *x*, *y*, and the *numerical value* of k.

OPTIONAL: TWO-POINT FORMULA

A two-point formula is one that has variables for two ordered pairs, x_1, y_1, and x_2, y_2. To derive a two-point formula, a constant, *k*, is first obtained for a direct or inverse formula. All ordered pairs for a particular relationship should have the same *k* value. For example, in a direct relationship ($y = k{\cdot}x$), first solve for *k* ($k = y/x$). Thus $y_1/x_1 = k$, and $y_2/x_2 = k$. Since both *k* values are the same:

$$\frac{y_1}{x_1} = \frac{y_2}{x_2}$$

In another example of an inverse square relationship, $y = k/x^2$ or $k = y{\cdot}x^2$. If $k = y_1{\cdot}x_1^2$ and $k = y_2{\cdot}x_2^2$, then:

$$y_1{\cdot}x_1^2 = y_2{\cdot}x_2^2$$

Derive a two-point formula for each of the three problems you have been assigned. Show the final answer in the space provided in the Calculation Table.

DATA AND CALCULATIONS TABLE

Problem Number _____	
X	Y
n = _____	

Problem Number _____	
X	Y
n = _____	

Problem Number _____	
X	Y
n = _____	

Problem Number	Equation (using x, y, & k)	Solve for "k" (find the value of k for two data pairs)	Final Equation (x, y, and value of k)
example	$y = k/x^2$	$k = y \cdot x^2$ $k = (4)(2)^2 = 16$ $k = (1)(4)^2 = 16$	$y = 16/x^2$

TEACHER INFORMATION

Find the Relationship:
An Exercise in Graphing Analysis

1. The student pages with complete instructions for data-collection using LabQuest App, Logger *Pro* (computers), EasyData or DataMate (calculators), and DataPro (Palm handhelds) can be found on the CD that accompanies this book. See *Appendix A* for more information.

2. This exercise is very useful prior to covering the gas laws and their accompanying experiments in this book. Depending on the number of computers available to the teacher, this exercise can either be completed during a regular class period or can be done on an individual basis over a period of a week. Students will become quite familiar with many analysis features of the Logger *Pro* program.

3. There are 30 problems provided on the following page. Teachers may make several photocopies of the page, and cut them up into individual problems. These may be placed in three stacks, one with *direct* relationships (#1-15), one with *inverse* relationships (#16-30), and a third with a mixture of problems. This way, each student should get a variety of types and difficulties of problems. An answer key is provided. Problems are arranged in rows with similar formulas: Row 1: $y = k \cdot x$; Row 2: $y = k \cdot x^2$; Row 3: $y = k \cdot x^3$; Row 4: $y = k/x$; Row 5: $y = k/x^2$; Row 6: $y = k/x^3$.

4. If the **XY Input** file is not on your calculator, it can be downloaded from the Vernier web site, www.vernier.com/easy/easydata.html. See *Appendix B* for information on transferring the file to your calculator.

5. To do this experiment, your TI graphing calculator should not be connected to a data-collection interface.

STUDENT PROBLEMS

1.

x	y
0.6	.198
0.8	.264
1.5	.495
2.0	.660
2.5	.825

7.

x	y
0.20	.290
0.25	.363
0.30	.435
0.45	.653
0.60	.870

13.

x	y
1.5	1.13
2.0	1.50
2.5	1.88
3.5	2.63
5.0	3.75

19.

x	y
45	405
37	333
16	144
4	36
64	576

25.

x	y
89	8.72
321	31.5
47	4.61
213	20.9
436	42.7

2.

x	y
32	819
13	135
43	1479
8	51.2
24	461

8.

x	y
0.7	16.7
1.2	49.0
1.8	110
4.5	689
2.5	213

14.

x	y
2	8
8	128
6	72
5	50
3	18

20.

x	y
4	8.0
3	4.5
7	24.5
5	12.5
8	32.0

26.

x	y
0.1	.002
0.2	.008
0.3	.018
0.5	.050
0.7	.098

3.

x	y
1	2
2	16
3	54
4	128
5	250

9.

x	y
0.4	.192
0.6	.648
0.2	.024
0.8	1.54
0.9	2.19

15.

x	y
0.1	.0005
0.5	.0625
0.3	.0135
0.7	.1715
1.0	.5000

21.

x	y
7	82.3
12	415
10	240
15	810
3	6.48

27.

x	y
0.5	2.25
0.8	9.22
1.1	24.0
1.9	123.5
0.2	0.144

4.

x	y
2.0	1.5
2.5	1.2
1.5	2.0
3.0	1.0
5.0	0.6

10.

x	y
2.5	10.00
1.5	16.67
12.0	2.083
19.0	1.316
5.0	5.000

16.

x	y
6	2.50
9	1.67
10	1.50
12	1.25
20	0.75

22.

x	y
14	5.36
25	3.00
19	3.95
36	2.08
48	1.56

28

x	y
5	0.16
2	0.40
8	0.10
15	0.053
0.5	1.60

5.

x	y
2.0	1.00
3.0	0.444
5.0	0.160
1.5	1.78
8.0	.0625

11.

x	y
5.0	0.84
2.0	5.25
3.0	2.33
4.8	0.91
7.5	0.373

17.

x	y
1.5	5.33
2.0	3.00
5.0	0.48
3.8	0.83
6.0	0.333

23.

x	y
5	5.80
11	1.20
15	0.644
24	0.252
30	0.161

29.

x	y
0.5	3.00
0.8	1.17
1.3	0.444
2.0	0.188
3.0	.0833

6.

x	y
11	0.236
17	.0639
6	1.454
28	.0143
20	.0393

12.

x	y
0.85	1.04
1.5	.190
0.5	5.12
2.0	.080
3.0	.0237

18.

x	y
0.4	78.1
0.9	6.85
0.7	14.6
1.2	2.89
0.6	23.2

24.

x	y
4.5	.154
5	.112
7	.041
3	.519
8	.0273

30.

x	y
2	3.00
3	0.889
5	0.192
7	0.070
9	0.033

ANSWERS TO PROBLEMS

1. $y = 0.33 \cdot x$
2. $y = 0.8 \cdot x^2$
3. $y = 2 \cdot x^3$
4. $y = 3/x$
5. $y = 4/x^2$
6. $y = 314/x^3$
7. $y = 1.45 \cdot x$
8. $y = 34 \cdot x^2$
9. $y = 3 \cdot x^3$
10. $y = 25/x$
11. $y = 21/x^2$
12. $y = 0.64/x^3$
13. $y = 0.75 \cdot x$
14. $y = 2 \cdot x^2$
15. $y = 0.5 \cdot x^3$
16. $y = 15/x$
17. $y = 12/x^2$
18. $y = 5/x^3$
19. $y = 9 \cdot x$
20. $y = 0.5 \cdot x^2$
21. $y = 0.24 \cdot x^3$
22. $y = 75/x$
23. $y = 145/x^2$
24. $y = 14/x^3$
25. $y = 0.098 \cdot x$
26. $y = 0.2 \cdot x^2$
27. $y = 18 \cdot x^3$
28. $y = 0.8/x$
29. $y = 0.75/x^2$
30. $y = 24/x^3$

SAMPLE PROBLEM

x	2.0	3.0	5.0	7.0	9.0
y	3.00	0.889	0.192	0.070	0.033

Solution to the problem: $y = 24/x^3$

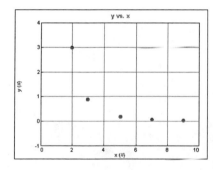

y vs. x

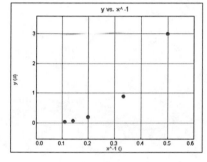

y vs. 1/x

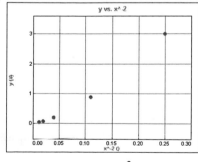

y vs. 1/x^2

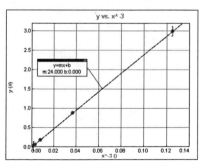

y vs. 1/x^3

Boyle's Law: Pressure-Volume Relationship in Gases

The primary objective of this experiment is to determine the relationship between the pressure and volume of a confined gas. The gas we use will be air, and it will be confined in a syringe connected to a Gas Pressure Sensor (see Figure 1). When the volume of the syringe is changed by moving the piston, a change occurs in the pressure exerted by the confined gas. This pressure change will be monitored using a Gas Pressure Sensor. It is assumed that temperature will be constant throughout the experiment. Pressure and volume data pairs will be collected during this experiment and then analyzed. From the data and graph, you should be able to determine what kind of mathematical relationship exists between the pressure and volume of the confined gas. Historically, this relationship was first established by Robert Boyle in 1662 and has since been known as Boyle's law.

OBJECTIVES

In this experiment, you will

- Use a Gas Pressure Sensor and a gas syringe to measure the pressure of an air sample at several different volumes.
- Determine the relationship between pressure and volume of the gas.
- Describe the relationship between gas pressure and volume in a mathematical equation.
- Use the results to predict the pressure at other volumes.

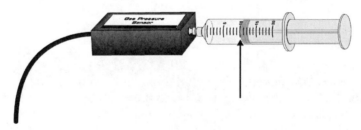

Figure 1

MATERIALS

computer
Vernier computer interface
LoggerPro

Vernier Gas Pressure Sensor
20 mL gas syringe

PROCEDURE

1. Prepare the Gas Pressure Sensor and an air sample for data collection.

   a. Plug the Gas Pressure Sensor into Channel 1 of the computer interface.

   b. With the 20 mL syringe disconnected from the Gas Pressure Sensor, move the piston of the syringe until the front edge of the inside black ring (indicated by the arrow in Figure 2) is positioned at the 10.0 mL mark.

   c. Attach the 20 mL syringe to the valve of the Gas Pressure Sensor.

2. Prepare the computer for data collection by opening the file "06 Boyle's Law" from the *Chemistry with Vernier* folder of Logger*Pro*.

3. To obtain the best data possible, you will need to correct the volume readings from the syringe. Look at the syringe; its scale reports its own internal volume. However, that volume is not the total volume of trapped air in your system since there is a little bit of space inside the pressure sensor.

   To account for the extra volume in the system, you will need to add 0.8 mL to your syringe readings. For example, with a 5.0 mL syringe volume, the total volume would be 5.8 mL. *It is this total volume that you will need for the analysis.*

4. Click ▶ Collect to begin data collection.

5. Collect the pressure *vs.* volume data. It is best for one person to take care of the gas syringe and for another to operate the computer.

   a. Move the piston to position the front edge of the inside black ring (see Figure 2) at the 5.0 mL line on the syringe. Hold the piston firmly in this position until the pressure value stabilizes.

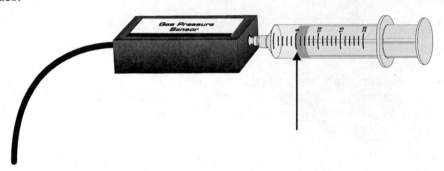

Figure 2

   b. When the pressure reading has stabilized, click ⊛ Keep . (The person holding the syringe can relax after ⊛ Keep is clicked.) Type in the total gas volume (in this case, 5.8 mL) in the edit box. Remember, you are adding 0.8 mL to the volume of the syringe for the total volume. Press the ENTER key to keep this data pair. Note: You can choose to redo a point by pressing the ESC key (after clicking ⊛ Keep but before entering a value).
   c. Move the piston to the 7.0 mL line. When the pressure reading has stabilized, click ⊛ Keep and type in the total volume, 7.8 mL.
   d. Continue this procedure for syringe volumes of 9.0, 11.0, 13.0, 15.0, 17.0, and 19.0 mL.
   e. Click ■ Stop when you have finished collecting data.

6. In your data table, record the pressure and volume data pairs displayed in the table (or, if directed by your instructor, print a copy of the table).

7. Examine the graph of pressure *vs.* volume. Based on this graph, decide what kind of mathematical relationship you think exists between these two variables, direct or inverse. To see if you made the right choice:

   a. Click the Curve Fit button, ⊡.
   b. Choose Variable Power ($y = Ax^n$) from the list at the lower left. Enter the power value, n, in the Power edit box that represents the relationship shown in the graph (e.g., type **1** if direct, **–1** if inverse). Click Try Fit .

c. A best-fit curve will be displayed on the graph. If you made the correct choice, the curve should match up well with the points. If the curve does not match up well, try a different exponent and click [Try Fit] again. When the curve has a good fit with the data points, then click [OK].

8. Once you have confirmed that the graph represents either a direct or inverse relationship, print a copy of the graph, with the graph of pressure *vs.* volume and its best-fit curve displayed.

9. With the best-fit curve still displayed, proceed directly to the Processing the Data section.

DATA AND CALCULATIONS

Volume (mL)	Pressure (kPa)	Constant, k (P/V or $P \cdot V$)

PROCESSING THE DATA

1. With the best-fit curve still displayed, choose Interpolate from the Analyze menu. A vertical cursor now appears on the graph. The cursor's volume and pressure coordinates are displayed in the floating box. Move the cursor along the regression line until the volume value is 5.0 mL. Note the corresponding pressure value. Now move the cursor until the volume value is doubled (10.0 mL). What does your data show happens to the pressure when the volume is *doubled*? Show the pressure values in your answer.

2. Using the same technique as in Question 1, what does your data show happens to the pressure if the volume is *halved* from 20.0 mL to 10.0 mL? Show the pressure values in your answer.

3. Using the same technique as in Question 1, what does your data show happens to the pressure if the volume is *tripled* from 5.0 mL to 15.0 mL? Show the pressure values in your answer.

4. From your answers to the first three questions *and* the shape of the curve in the plot of pressure *vs.* volume, do you think the relationship between the pressure and volume of a confined gas is direct or inverse? Explain your answer.

5. Based on your data, what would you expect the pressure to be if the volume of the syringe was increased to 40.0 mL? Explain or show work to support your answer.

6. Based on your data, what would you expect the pressure to be if the volume of the syringe was decreased to 2.5 mL? Explain or show work to support your answer.

7. What experimental factors are assumed to be constant in this experiment?

8. One way to determine if a relationship is inverse or direct is to find a proportionality constant, *k*, from the data. If this relationship is direct, $k = P/V$. If it is inverse, $k = P \cdot V$. Based on your answer to Question 4, choose one of these formulas and calculate k for the seven ordered pairs in your data table (divide or multiply the *P* and *V* values). Show the answers in the third column of the Data and Calculations table.

9. How *constant* were the values for *k* you obtained in Question 8? Good data may show some minor variation, but the values for *k* should be relatively constant.

10. Using *P, V,* and *k*, write an equation representing Boyle's law. Write a verbal statement that correctly expresses Boyle's law.

EXTENSION

1. To confirm the type of relationship that exists between pressure and volume, a graph of pressure versus the *reciprocal of volume* (1/volume or volume^{-1}) may also be plotted. To do this using Logger*Pro*, it is necessary to create a new column of data, reciprocal of volume, based on your original volume data.

   a. Remove the Curve Fit box from the graph by clicking on its upper-left corner.
   b. Choose New Calculated Column from the Data menu.
   c. Enter "1/Volume" as the Name, "1/V" as the Short Name, and "1/mL" as the Unit. Enter the correct formula for the column (1/volume) into the Equation edit box. To do this, type in "1" and "/". Then select "Volume" from the Variables list. In the Equation edit box, you should now see displayed: 1/"Volume". Click ☐ Done ☐.
   d. Click on the horizontal-axis label, select "1/Volume" to be displayed on the horizontal axis.

2. Decide if the new relationship is direct or inverse and change the formula in the Fit menu accordingly.

   a. Click the Curve Fit button, ☒.
   b. Choose Variable Power from the list at the lower left. Enter the value of the power in the edit box that represents the relationship shown in the graph (e.g., type "1" if direct, "–1" if inverse). Click ☐ Try Fit ☐.
   c. A best-fit curve will be displayed on the graph. If you made the correct choice, the curve should match up well with the points. If the curve does not match up well, try a different exponent and click ☐ Try Fit ☐ again. When the curve has a good fit with the data points, then click ☐ OK ☐.

3. If the relationship between *P* and *V* is an inverse relationship, the plot of *P* vs. 1/V should be direct; that is, the curve should be linear and pass through (or near) your data points. Examine your graph to see if this is true for your data.

4. Print a copy of the graph.

TEACHER INFORMATION

Boyle's Law: Pressure-Volume Relationship in Gases

1. The student pages with complete instructions for data-collection using LabQuest App, Logger *Pro* (computers), EasyData or DataMate (calculators), and DataPro (Palm handhelds) can be found on the CD that accompanies this book. See *Appendix A* for more information.

2. This experiment is written for the Gas Pressure Sensor. The default calibration for this experiment has units of kPa (kilopascals). You can use other units (mm Hg, atm, or psi) by changing them in the data-collection software.

3. In order to save time, you may prefer to do Step 1 of the student procedure prior to the start of class.

4. You should never have to perform a new calibration when using the Gas Pressure Sensor in pressure experiments. Simply use the stored calibration.

5. As explained in the student procedures, this experiment is written to compensate for the small inside volume of the white stem that leads to the inside of the Gas Pressure Sensor. The volume of this space is about 0.8 mL. This means that when students enter a volume of 5.0 mL (as read on the syringe), the volume is really about 5.8 mL. To compensate for this error, the students are instructed add 0.8 mL to each of the volumes they enter. By doing this, they will get better results for the value of the exponent, *b*, in Step 7.

6. Question 8 in the Processing the Data section asks the students to calculate a proportionality constant, *k*, using the equation, $k = P \cdot V$. Your students can do this manually, or you could have them create a calculated column using Logger *Pro*. To do the latter, import the data to Logger *Pro* if it is not already in there and give them these instructions.

    a. Choose New Calculated Column from the Data menu.

    b. Enter the Name, Short Name, and Units for the new column.

    c. Enter the formula into the Equation edit box. Use the Variables drop-down list to choose Volume and Pressure. Depending on your choice of formula, type the division (/) or multiplication (*) character in at the appropriate point in the formula.

    d. Click $\boxed{\text{Done}}$.

ANSWERS TO QUESTIONS

1. When the volume was doubled, the pressure was halved (pressure went from 204.6 kPa to 103.3 kPa).

2. When the volume was halved, the pressure doubled (pressure went from 50.7 kPa to 103.3 kPa).

3. The pressure is reduced by a factor of 1/3 (pressure went from 204.6 kPa to 69.9 kPa).

4. From the data, the relationship appears to be inverse. When pressure data increases, volume data seems to decrease proportionally. The shape of the pressure-volume plot appears to be a simple inverse relationship.

5. If the volume is increased to 40.0 mL, one would expect the pressure to be 1/2 of what it was at 20.0 mL. This would be a pressure of approximately 25 kPa.

6. If the volume were reduced to 2.5 mL, one would expect the pressure to be double what it was at 5.0 mL. This would be a pressure of approximately 400 kPa.

7. The temperature and the number of molecules in the gas sample are assumed to be constant.

8. The correct formula for an inverse relationship is: $k = P \cdot V$. For k values, see the third column of the sample results on this page (1027 kPa·mL is the average value for the constant, k).

9. Values were quite constant, with a very small deviation.

10. The equation representing Boyle's law is: $k = P \cdot V$. The pressure of a confined gas varies inversely with the volume of the gas if the temperature of the sample remains constant.

SAMPLE RESULTS

Volume (mL)	Pressure (kPa)	Constant, k (kPa·mL)
5.8	175.9	1020
7.8	131.4	1025
9.8	105.1	1030
11.8	87.0	1027
13.8	74.4	1027
15.8	65.1	1029
17.8	57.6	1025
19.8	52.0	1030

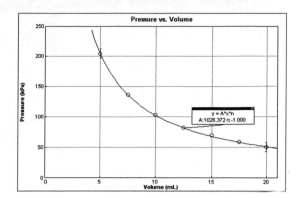

Pressure vs. Volume

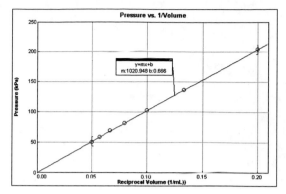

Pressure vs. Reciprocal of Volume

Pressure -Temperature Relationship in Gases

Gases are made up of molecules that are in constant motion and exert pressure when they collide with the walls of their container. The velocity and the number of collisions of these molecules are affected when the temperature of the gas increases or decreases. In this experiment, you will study the relationship between the temperature of a gas sample and the pressure it exerts. Using the apparatus shown in Figure 1, you will place an Erlenmeyer flask containing an air sample in water baths of varying temperature. Pressure will be monitored with a Gas Pressure Sensor and temperature will be monitored using a Temperature Probe. The volume of the gas sample and the number of molecules it contains will be kept constant. Pressure and temperature data pairs will be collected during the experiment and then analyzed. From the data and graph, you will determine what kind of mathematical relationship exists between the pressure and absolute temperature of a confined gas. You may also do the extension exercise and use your data to find a value for absolute zero on the Celsius temperature scale.

OBJECTIVES

In this experiment, you will

- Study the relationship between the temperature of a gas sample and the pressure it exerts.
- Determine from the data and graph, the mathematical relationship between the pressure and absolute temperature of a confined gas.
- Find a value for absolute zero on the Celsius temperature scale.

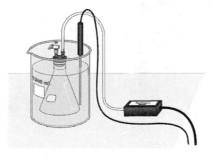

Figure 1

MATERIALS

computer	125 mL Erlenmeyer flask
Vernier computer interface	ring stand
Logger *Pro*	utility clamp
Vernier Gas Pressure Sensor	hot plate
Vernier Temperature Probe	four 1 liter beakers
plastic tubing with two connectors	glove or cloth
rubber stopper assembly	ice

PROCEDURE

1. Obtain and wear goggles.

2. Prepare a boiling-water bath. Put about 800 mL of hot tap water into a l L beaker and place it on a hot plate. Turn the hot plate to a high setting.

3. Prepare an ice-water bath. Put about 700 mL of cold tap water into a second 1 L beaker and add ice.

4. Put about 800 mL of room-temperature water into a third 1 L beaker.

5. Put about 800 mL of hot tap water into a fourth 1 L beaker.

6. Prepare the Temperature Probe and Gas Pressure Sensor for data collection.

    a. Plug the Gas Pressure Sensor into CH1 and the Temperature Probe into CH2 of the computer interface.

    b. Obtain a rubber-stopper assembly with a piece of heavy-wall plastic tubing connected to one of its two valves. Attach the connector at the free end of the plastic tubing to the open stem of the Gas Pressure Sensor with a clockwise turn. Leave its two-way valve on the rubber stopper open (lined up with the valve stem as shown in Figure 2) until Step 9.

    c. Insert the rubber-stopper assembly into a 125 mL Erlenmeyer flask. **Important:** Twist the stopper into the neck of the flask to ensure a tight fit.

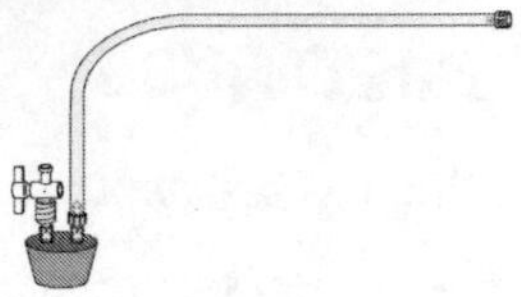

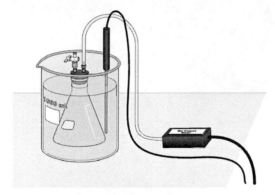

Figure 2

    d. Close the 2-way valve above the rubber stopper—do this by turning the valve handle so it is perpendicular with the valve stem itself (as shown in Figure 3). The air sample to be studied is now confined in the flask.

Figure 3

7. Prepare the computer for data collection by opening the file "07 Pressure-Temperature" from the *Chemistry with Vernier* folder of Logger*Pro*.

8. Click ▶ Collect to begin data collection.

9. Collect pressure *vs.* temperature data for your gas sample:

    a. Place the flask into the ice-water bath. Make sure the entire flask is covered (see Figure 4). Stir.

    b. Place the temperature probe into the ice-water bath.

    c. When the pressure and temperature readings displayed in the meter stabilize, click ⊛ Keep. You have now saved the first pressure-temperature data pair.

10. Repeat the Step-9 procedure using the room-temperature bath.

11. Repeat the Step-9 procedure using the hot-water bath.

12. Use a ring stand and utility clamp to suspend the temperature probe in the boiling-water bath. To keep from burning your hand, hold the tubing of the flask using a glove or a cloth. After the temperature probe has been in the boiling water for a few seconds, place the flask into the boiling-water bath and repeat the Step-9 procedure. Remove the flask and the temperature probe after you have clicked ⊛ Keep. **CAUTION:** *Do not burn yourself or the probe wires with the hot plate.*

13. Click ■ Stop when you have finished collecting data. Turn off the hot plate. Record the pressure and temperature values in your data table, or, if directed by your instructor, print a copy of the table.

14. Examine your graph of pressure *vs.* temperature (°C). In order to determine if the relationship between pressure and temperature is direct or inverse, you must use an absolute temperature scale; that is, a temperature scale whose 0° point corresponds to absolute zero. We will use the Kelvin absolute temperature scale. Instead of manually adding 273 to each of the Celsius temperatures to obtain Kelvin values, you can create a new data column for Kelvin temperature.

    a. Choose New Calculated Column from the Data menu.

    b. Enter "Temp Kelvin" as the Name, "T Kelvin" as the Short Name, and "K" as the Unit. Enter the correct formula for the column into the Equation edit box. Type in "273+". Then select "Temperature" from the Variables list. In the Equation edit box, you should now see displayed: 273+"Temperature". Click Done.

    c. Click on the horizontal axis label and select "Temp Kelvin" to be displayed on the horizontal axis.

15. Decide if your graph of pressure *vs.* temperature (K) represents a direct or inverse relationship:

    a. Click the Curve Fit button, ⤳.

    b. Choose your mathematical relationship from the list at the lower left. If you think the relationship is linear (or direct), use Linear. If you think the relationship represents a power, use Power. Click Try Fit.

    c. A best-fit curve will be displayed on the graph. If you made the correct choice, the curve should match up well with the points. If the curve does not match up well, try a different mathematical function and click Try Fit again. When the curve has a good fit with the data points, then click OK.

    d. Autoscale both axes from zero by double-clicking in the center of the graph to view Graph Options. Click the Axis Options tab, and select Autoscale from 0 for both axes.

16. Print a copy of the graph of pressure *vs.* temperature (K). The regression line should still be displayed on the graph. Enter your name(s) and the number of copies you want to print.

PROCESSING THE DATA

1. In order to perform this experiment, what two experimental factors were kept constant?

2. Based on the data and graph that you obtained for this experiment, express in words the relationship between gas pressure and temperature.

3. Explain this relationship using the concepts of molecular velocity and collisions of molecules.

4. Write an equation to express the relationship between pressure and temperature (K). Use the symbols P, T, and k.

5. One way to determine if a relationship is inverse or direct is to find a proportionality constant, k, from the data. If this relationship is direct, $k = P/T$. If it is inverse, $k = P \cdot T$. Based on your answer to Question 4, choose one of these formulas and calculate k for the four ordered pairs in your data table (divide or multiply the P and T values). Show the answer in the fourth column of the Data and Calculations table. How "constant" were your values?

6. According to this experiment, what should happen to the pressure of a gas if the Kelvin temperature is doubled? Check this assumption by finding the pressure at $-73°C$ (200 K) and at $127°C$ (400 K) on your graph of pressure versus temperature. How do these two pressure values compare?

DATA AND CALCULATIONS

Pressure (kPa)	Temperature (°C)	Temperature (K)	Constant, k (P/T or $P \cdot T$)

EXTENSION

The data that you have collected can also be used to determine the value for absolute zero on the Celsius temperature scale. Instead of plotting pressure versus Kelvin temperature like we did above, this time you will plot Celsius temperature on the y-axis and pressure on the x-axis. Since absolute zero is the temperature at which the pressure theoretically becomes equal to zero, the temperature where the regression line (the extension of the temperature-pressure curve) intercepts the y-axis should be the Celsius temperature value for absolute zero. You can use the data you collected in this experiment to determine a value for absolute zero.

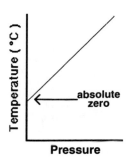

1. Remove the curve fit box on the graph by clicking on its upper-left corner.

2. Click on the vertical-axis label and select "Temperature" to plot the Celsius temperature. In the same way, select "Pressure" to be displayed on the horizontal axis.

3. Rescale the temperature axis from a minimum of –300°C to a maximum of 200°C. This may be done by clicking on the minimum or maximum value displayed on the graph axis and editing them. The pressure axis should be scaled from 0 kPa to 150 kPa.

4. Click the Linear Fit button, ⬚. A best-fit linear regression curve will be shown for the four data points. The equation for the regression line will be displayed in a box on the graph, in the form $y = mx + b$. The numerical value for b is the y-intercept and represents the Celsius value for absolute zero.

5. Print the graph of temperature (°C) *vs.* pressure, with the regression line and its regression statistics still displayed. Enter your name(s) and the number of copies you want to print. Clearly label the position and value of absolute zero on the printed graph.

Pressure -Temperature Relationship in Gases

1. The student pages with complete instructions for data-collection using LabQuest App, Logger *Pro* (computers), EasyData or DataMate (calculators), and DataPro (Palm handhelds) can be found on the CD that accompanies this book. See *Appendix A* for more information.

2. The boiling-water bath is not absolutely necessary. You can, for safety reasons, instruct your students to use a highest temperature that is somewhat lower than 100°C and still get good results.

3. As a safer alternative to students handling the boiling water bath, you can move a lab cart with a hot plate and water bath to different lab stations. You can plug the hot plate in at each lab station and assist students with this trial.

4. All of the pressure valves, tubing, and connectors used in this experiment are included with Vernier Pressure Sensors shipped after February 15, 1998. These accessories are also helpful when performing vapor pressure experiments such as Experiment 10 in this manual, *Vapor Pressure of Liquids*.

   If you purchased your pressure sensor at an earlier date, or need replacement parts, Vernier has a Pressure Sensor Accessories Kit (PS-ACC) that includes all of the parts shown here for doing pressure-temperature experiments (except the Erlenmeyer flask). Using this kit allows for easy assembly of a completely airtight system. The kit includes the following parts:

   - two ribbed, tapered valve connectors inserted into a No. 5 rubber stopper
   - one ribbed, tapered valve connectors inserted into a No. 1 rubber stopper
   - two Luer-lock connectors connected to either end of a piece of plastic tubing
   - one two-way valve
   - one 20 mL syringe
   - two tubing clamps for transpiration experiments

5. Connect the piece of plastic tubing to the shorter valve connector on the rubber stopper prior to the start of class (as shown here). Threaded Luer-lock connectors are attached to both ends of the plastic tubing. Connect one of these to the valve stem with a 1/2 clockwise turn.

6. The stored temperature calibration for the Stainless Steel Temperature Probe or Direct-Connect Temperature Probe works well for this experiment.

   The stored pressure calibration for the Gas Pressure Sensor also works well for this experiment.

SAMPLE RESULTS

Pressure (kPa)	Temperature (°C)	Temperature (K)	Contstant, K (P/T)
96.5	0.0	273.0	0.354
102.3	19.9	292.9	0.349
110.7	44.2	317.2	0.349
130.8	99.7	372.7	0.351

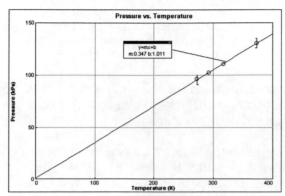

Pressure vs. temperature (K)

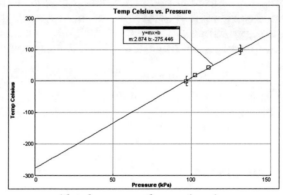

Absolute zero determination

ANSWERS TO QUESTIONS

1. Volume of the gas and number of molecules were kept constant.

2. Pressure varies directly with the absolute temperature, if the volume and number of gas molecules are kept constant.

3. An increase in temperature causes molecules to move at a higher velocity. Faster-moving molecules have more collisions with the walls of the container, resulting in a higher pressure.

4. $P/T = k$ (or $T/P = k$)

5. P/T was a constant. The average value was 0.351 kPa/K (see values in the sample results).

6. Pressure is expected to double if absolute temperature doubles. The pressure at 200 K is ~ 70 kPa and the pressure at 400 K is ~140 kPa. The second pressure is double that of the first.

Fractional Distillation

An example of a simple distillation is the separation of a solution of salt and water into two separate pure substances. When the salt water solution is heated to boiling, water vapor from the mixture reaches the condenser (see Figure 1) and the cold water circulating around the inside tube causes condensation of water vapor into droplets of liquid water. The liquid water is then collected at the lower end of the condenser. The non-volatile salt remains in the flask.

In this experiment, the initial mixture you distill contains two volatile liquids: ethanol and water. In this distillation, *both* of the liquids will evaporate from the boiling solution. Ethanol and water have normal boiling temperatures of 79°C and 100°C, respectively. One objective of the experiment is to observe what happens when a liquid-liquid mixture is heated and allowed to boil over a period of time. Throughout the distillation, volumes of distillate, called *fractions*, will be collected. The percent composition of ethanol and water in each fraction will be determined from its density. Water has a density of 1.00 g/cm^3 (at 20°C) and ethanol has a density of 0.79 g/cm^3 (at 20°C). The fractions you collect will have densities in this range.

OBJECTIVES

In this experiment, you will

- Observe what happens when a liquid-liquid mixture is heated and allowed to boil over a period of time.
- Determine percent composition of ethanol and water in the fraction from its density.

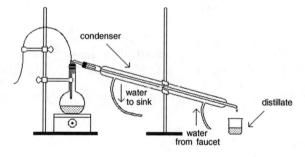

Figure 1

MATERIALS

computer	three 100 mL beakers
Vernier computer interface	500 mL flask
Logger *Pro*	2 ring stands
Temperature Probe	3 utility clamps
condenser with two hoses	50 mL graduated cylinder
hot plate	100 mL graduated cylinder
ethanol, C$_2$H$_5$OH, denatured (94-96%)	3–4 boiling chips
water	2-hole stopper
bent-glass tubing	

PROCEDURE

1. Obtain and wear goggles! **CAUTION:** The compounds used in this experiment are flammable and poisonous. Avoid inhaling their vapors. Avoid contacting them with your skin or clothing. Be sure there are no open flames in the lab during this experiment. Notify your teacher immediately if an accident occurs.

2. Assemble the distillation apparatus as shown in Figure 1. Do not begin heating until your teacher has checked your apparatus.

3. Use a 100 mL graduated cylinder to obtain 60 mL of ethanol. Pour the ethanol into the flask. Put 60 mL of tap water into the same flask. Add 3–4 boiling chips to ensure the formation of many small bubbles during the boiling.

4. Make sure the stopper fitting the probe into the flask and the stopper fitting the glass bend into the condenser are tightly in place. Turn on the cold tap water so that it slowly flows up through the condenser.

5. Use a third utility clamp, fitted at the top of the ring stand, to secure the Temperature Probe wire as far as possible from the hot plate.

6. Connect the probe to the computer interface. Prepare the computer for data collection by opening the "08 Fractional Distillation" file from the *Chemistry with Vernier* folder of Logger*Pro*.

7. Label the three 100 mL beakers 1–3. Put Beaker 1 in place at the end of the condenser. Have your teacher check your set-up, and then turn the hot plate on to its maximum setting. **CAUTION:** Do not burn yourself or the probe wire with the hot plate.

8. Click ▶ Collect to begin data collection. When the temperature reaches 50°C, turn the hot plate down to a medium setting.

9. When 30 mL of liquid (distillate) have been collected in Beaker 1, remove it and insert Beaker 2 in its place. This 30 mL portion is called Fraction 1. Set it aside until Step 12.

10. After you have collected 30.0 mL of Fraction 2 in Beaker 2, quickly replace it with beaker #3. Set Fraction 2 aside until Step 13.

11. After you have collected 30.0 mL of Fraction 3 in Beaker 3, click ■ Stop to end data collection. Turn off the hot plate.

12. Determine and record the mass of a clean, dry 50 mL graduated cylinder. Pour the contents of Fraction 1 into the graduated cylinder. Read and record its volume, to the nearest 0.1 mL. Then determine and record the mass of the distillate plus the graduated cylinder. Discard the distillate as directed by your teacher.

13. Pour the contents of Fraction 2 into the 50 mL graduated cylinder. Read and record its volume, to the nearest 0.1 mL. Then determine and record the mass of the distillate plus the graduated cylinder. Discard the distillate as directed by your teacher.

14. Repeat the Step-13 procedure for Fraction 3.

15. Print a graph of temperature *vs.* time for the distillation. Enter your name(s) and the number of copies of the graph you want.

16. Find the initial boiling temperature of the mixture (the point where the rapidly rising initial temperature ends, and the slow increase in temperature begins). Click the Examine button,

☒. As you move the mouse pointer along the curve, examine the data values in the display box on the graph. When you determine the initial boiling temperature, label this value on your printed copy of the graph.

DENSITY OF ETHANOL AND WATER MIXTURES (20°C)

Ethanol (%)	Density (g/cm³)	Ethanol (%)	Density (g/cm³)	Ethanol (%)	Density (g/cm³)
0	0.998	34	0.947	68	0.872
2	0.995	36	0.943	70	0.868
4	0.991	38	0.939	72	0.863
6	0.988	40	0.935	74	0.858
8	0.985	42	0.931	76	0.853
10	0.982	44	0.927	78	0.848
12	0.979	46	0.923	80	0.843
14	0.977	48	0.918	82	0.839
16	0.974	50	0.913	84	0.834
18	0.971	52	0.909	86	0.828
20	0.969	54	0.905	88	0.823
22	0.966	56	0.900	90	0.818
24	0.964	58	0.896	92	0.813
26	0.960	60	0.891	94	0.807
28	0.957	62	0.887	96	0.801
30	0.954	64	0.882	98	0.795
32	0.950	66	0.877	100	0.789

PROCESSING THE DATA

1. After finding the mass of the distillate by subtracting the mass of the graduated cylinder from the mass of the graduated cylinder + distillate, calculate the density of each fraction using the formula: density = mass/volume.

2. Using the density table above, determine the *% ethanol* corresponding to the density of each fraction. Record these values.

3. Using the values of % ethanol obtained in the previous step, determine the *% water* for each fraction.

4. What is the primary component of the first fraction you collected? Explain why it is not pure.

5. Did the density of the fractions increase or decrease as the experiment progressed? Explain.

6. What happened to the % of ethanol in the collected fractions as the experiment progressed? What happened to the % of water?

7. What could be done to subsequently increase the purity of the ethanol (reduce the water) in the first fraction? Explain.

8. In Step 16 of the procedure, you found (and recorded on your graph) the initial boiling temperature of the mixture. Is this value lower or higher than the normal boiling temperature of pure ethanol (79°C)?

DATA AND CALCULATIONS

	Fraction 1	Fraction 2	Fraction 3
Mass of distillate plus graduated cylinder	g	g	g
Mass of graduated cylinder	g	g	g
Mass of distillate	g	g	g
Volume of distillate	cm^3	cm^3	cm^3
Initial boiling temperature	°C		

Density			
	g/cm^3	g/cm^3	g/cm^3
Percent ethanol	%	%	%
Percent water	%	%	%

OBSERVATIONS

TEACHER INFORMATION

Fractional Distillation

1. The student pages with complete instructions for data-collection using LabQuest App, Logger *Pro* (computers), EasyData or DataMate (calculators), and DataPro (Palm handhelds) can be found on the CD that accompanies this book. See *Appendix A* for more information.

2. The ethanol (ethyl alcohol) used in this lab can be relatively low-grade denatured alcohol (approximately 95%). **HAZARD ALERT:** Dangerous fire risk; flammable; addition of denaturant makes the product poisonous—it cannot be made non-poisonous; store in a dedicated flammables cabinet or safety cans. If a flammables cabinet or safety cans are not available, store in a Flinn *Saf-Stor*® Can. Hazard Code: C—Somewhat hazardous.

    The hazard information reference is: Flinn Scientific, Inc., *Chemical & Biological Catalog Reference Manual,*(800) 452-1261, www.flinnsci.com. See *Appendix D* of this book, *Chemistry with Vernier*, for more information.

3. The best heat source is a hot plate. It eliminates the hazard of having an open flame near ethanol vapor. A hot plate operates at lower temperatures, so the distillation takes place more slowly.

4. Equipment substitutions, such as the use of a burner and boiling flask, can also be made.

5. It is very important to check the apparatus of each student pair before they begin heating. Make sure the temperature probe cord is well away from the hot plate. Be sure all stoppers are firmly in place. Also, be certain students have water flowing through the condenser. Watch to make sure the heating does not take place too fast.

6. When using a Stainless Steel Temperature Probe, be sure to use a two-hole stopper with a holes small enough to provide a snug, airtight fit for the probe tip (or bore a smaller hole yourself). For airtight fits and low probability of temperature probe damage, we recommend the use of *Twistit* stoppers in this experiment. Be sure to use a lubricant when inserting or removing probes. Carefully remove the probes from the stopper on the same day of the experiment. Twistit stoppers may be purchased from Sargent-Welch Scientific Company, www.sargentwelch.com.

7. Because this experiment requires a full class period, you may want your students to set up the apparatus the day before.

ANSWERS TO QUESTIONS

4. Ethanol is the primary component of the first fraction. It is not pure because water molecules will also evaporate and condense in this fraction.

5. The density increased, due to the higher percentage of water in subsequent fractions. Water has a higher density (1.00 g/cm^3) than ethanol (0.79 g/cm^3).

6. The percentage of ethanol decreased and the percentage of water increased.

7. After the first fraction is collected, it can be re-distilled to increase its purity.

8. The initial boiling temperature is approximately 82°C, which is approximately 3°C higher than the normal boiling temperature of pure ethanol. Due to the presence of a second substance in this solution, the boiling temperature has been elevated.

SAMPLE RESULTS

	Fraction 1	Fraction 2	Fraction 3
Mass of distillate plus graduated cylinder	90.24 g	91.05 g	91.97 g
Mass of graduated cylinder	64.52 g	64.52 g	64.52 g
Mass of distillate	25.72 g	26.53 g	27.45 g
Volume of distillate	30.5 cm^3	30.5 cm^3	30.1 cm^3
Initial boiling temperature	82.0°C		

	Fraction 1	Fraction 2	Fraction 3
Density	$\dfrac{25.72 \text{ g}}{30.5 \text{ cm}^3} =$ 0.843 g/cm^3	$\dfrac{26.53 \text{ g}}{30.5 \text{ cm}^3} =$ 0.870 g/cm^3	$\dfrac{27.45 \text{ g}}{30.1 \text{ cm}^3} =$ 0.912 g/cm^3
Percent ethanol	79 %	69 %	50 %
Percent water	21 %	31 %	50 %

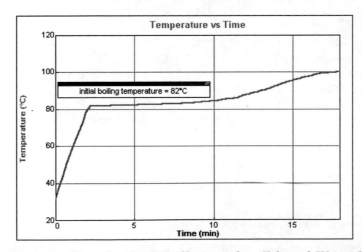

Temperature vs. Time for the Distillation of an Ethanol-Water Mixture

Evaporation and Intermolecular Attractions

In this experiment, Temperature Probes are placed in various liquids. Evaporation occurs when the probe is removed from the liquid's container. This evaporation is an endothermic process that results in a temperature decrease. The magnitude of a temperature decrease is, like viscosity and boiling temperature, related to the strength of intermolecular forces of attraction. In this experiment, you will study temperature changes caused by the evaporation of several liquids and relate the temperature changes to the strength of intermolecular forces of attraction. You will use the results to predict, and then measure, the temperature change for several other liquids.

You will encounter two types of organic compounds in this experiment—alkanes and alcohols. The two alkanes are n-pentane, C_5H_{12}, and n-hexane, C_6H_{14}. In addition to carbon and hydrogen atoms, alcohols also contain the -OH functional group. Methanol, CH_3OH, and ethanol, C_2H_5OH, are two of the alcohols that we will use in this experiment. You will examine the molecular structure of alkanes and alcohols for the presence and relative strength of two intermolecular forces—hydrogen bonding and dispersion forces.

OBJECTIVES

In this experiment, you will

- Study temperature changes caused by the evaporation of several liquids.
- Relate the temperature changes to the strength of intermolecular forces of attraction.

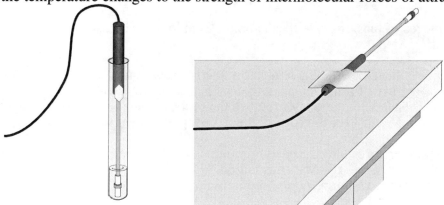

Figure 1

MATERIALS

computer	methanol (methyl alcohol)
Vernier computer interface	ethanol (ethyl alcohol)
Logger *Pro*	1-propanol
two Temperature Probes	1-butanol
6 pieces of filter paper (2.5 cm × 2.5 cm)	n-pentane
2 small rubber bands	n-hexane
masking tape	

PRE-LAB EXERCISE

Prior to doing the experiment, complete the Pre-Lab table. The name and formula are given for each compound. Draw a structural formula for a molecule of each compound. Then determine the molecular weight of each of the molecules. Dispersion forces exist between any two molecules, and generally increase as the molecular weight of the molecule increases. Next, examine each molecule for the presence of hydrogen bonding. Before hydrogen bonding can occur, a hydrogen atom must be bonded directly to an N, O, or F atom within the molecule. Tell whether or not each molecule has hydrogen-bonding capability.

PROCEDURE

1. Obtain and wear goggles! **CAUTION:** The compounds used in this experiment are flammable and poisonous. Avoid inhaling their vapors. Avoid contacting them with your skin or clothing. Be sure there are no open flames in the lab during this experiment. Notify your instructor immediately if an accident occurs.

2. Connect the probes to the computer interface. Prepare the computer for data collection by opening the file "09 Evaporation" from the *Chemistry with Vernier* folder.

3. Wrap Probe 1 and Probe 2 with square pieces of filter paper secured by small rubber bands as shown in Figure 1. Roll the filter paper around the probe tip in the shape of a cylinder. Hint: First slip the rubber band up on the probe, wrap the paper around the probe, and then finally slip the rubber band over the wrapped paper. The paper should be even with the probe end.

4. Stand Probe 1 in the ethanol container and Probe 2 in the 1-propanol container. Make sure the containers do not tip over.

5. Prepare 2 pieces of masking tape, each about 10 cm long, to be used to tape the probes in position during Step 6.

6. After the probes have been in the liquids for at least 30 seconds, begin data collection by clicking $\boxed{\text{▶ Collect}}$. Monitor the temperature for 15 seconds to establish the initial temperature of each liquid. Then simultaneously remove the probes from the liquids and tape them so the probe tips extend 5 cm over the edge of the table top as shown in Figure 1.

7. When both temperatures have reached minimums and have begun to increase, click $\boxed{\text{■ Stop}}$ to end data collection. Click the Statistics button, $\boxed{\text{STAT}}$, then click $\boxed{\text{OK}}$ to display a box for both probes. Record the maximum (t_1) and minimum (t_2) values for Temperature 1 (ethanol) and Temperature 2 (1-propanol).

8. For each liquid, subtract the minimum temperature from the maximum temperature to determine Δt, the temperature change during evaporation.

9. Roll the rubber band up the probe shaft and dispose of the filter paper as directed by your instructor.

10. Based on the Δt values you obtained for these two substances, plus information in the Pre-Lab exercise, *predict* the size of the Δt value for 1-butanol. Compare its hydrogen-bonding capability and molecular weight to those of ethanol and 1-propanol. Record your predicted Δt, then explain how you arrived at this answer in the space provided. Do the same for n-pentane. It is not important that you predict the exact Δt value; simply estimate a logical value that is higher, lower, or between the previous Δt values.

11. Test your prediction in Step 10 by repeating Steps 3-9 using 1-butanol for Probe 1 and n-pentane for Probe 2.

12. Based on the Δt values you have obtained for all four substances, plus information in the Pre-Lab exercise, predict the Δt values for methanol and n-hexane. Compare the hydrogen-bonding capability and molecular weight of methanol and n-hexane to those of the previous four liquids. Record your predicted Δt, then explain how you arrived at this answer in the space provided.

13. Test your prediction in Step 12 by repeating Steps 3–9, using methanol with Probe 1 and n-hexane with Probe 2.

PROCESSING THE DATA

1. Two of the liquids, n-pentane and 1-butanol, had nearly the same molecular weights, but significantly different Δt values. Explain the difference in Δt values of these substances, based on their intermolecular forces.

2. Which of the alcohols studied has the strongest intermolecular forces of attraction? The weakest intermolecular forces? Explain using the results of this experiment.

3. Which of the alkanes studied has the stronger intermolecular forces of attraction? The weaker intermolecular forces? Explain using the results of this experiment.

4. Plot a graph of Δt values of the four alcohols versus their respective molecular weights. Plot molecular weight on the horizontal axis and Δt on the vertical axis.

PRE-LAB

Substance	Formula	Structural Formulas	Molecular Weight	Hydrogen Bond (Yes or No)
ethanol	C_2H_5OH			
1-propanol	C_3H_7OH			
1-butanol	C_4H_9OH			
n-pentane	C_5H_{12}			
methanol	CH_3OH			
n-hexane	C_6H_{14}			

DATA TABLE

Substance	t_1 (°C)	t_2 (°C)	Δt (t_1-t_2) (°C)
ethanol			
1-propanol			
1-butanol			
n-pentane			
methanol			
n-hexane			

Predicted Δt (°C)	Explanation

TEACHER INFORMATION

Evaporation and Intermolecular Attractions

1. The student pages with complete instructions for data-collection using LabQuest App, Logger *Pro* (computers), EasyData or DataMate (calculators), and DataPro (Palm handhelds) can be found on the CD that accompanies this book. See *Appendix A* for more information.

2. This activity is ideally performed with a LabQuest, LabPro, or CBL 2 and two Stainless Steel Temperature Probes. It is possible to perform this activity with one Temperature Probe, but it will take additional time.

   If you use only one Temperature Probe, this experiment can still be completed in one class period. It is also possible to do four of the liquids during one class period, and the remaining two liquids the next day. This provides students with additional time to consider their predictions.

   Note: We do not recommend that you use the TI Temperature Probe that was shipped with the original CBL; our tests show that pentane liquid sometimes penetrates the seal on the tip of the TI-Temperature Probe.

3. We recommend wrapping the probes with paper as described in the procedure. Wrapped probes provide more uniform liquid amounts, and generally greater Δt values, than bare probes. Chromatography paper, filter paper, and various other paper types work well.

4. Snug-fitting rubber bands can be made by cutting short sections from a small rubber hose. Surgical tubing works well. Orthodontist's rubber bands are also a good size.

5. Other liquids can be substituted. Although it has a somewhat larger Δt, 2-propanol can be substituted for 1-propanol. Some petroleum ethers have a high percentage of hexane and can be used in its place. Other alkanes of relatively high purity, such as n-heptane or n-octane can be used. Water, with a Δt value of about 5°C, emphasizes the effect of hydrogen bonding on a low-molecular weight liquid. However, students might have difficulty comparing its hydrogen bonding capability with that of the alcohols used.

6. Sets of the liquids can be supplied in 13 × 100 mm test tubes stationed in stable test-tube racks. This method uses very small amounts of the liquids. Alternatively, the liquids can be supplied in sets of small bottles kept for future use. Adjust the level of the liquids in the containers so it will be above the top edge of the filter paper.

7. Because several of these liquids are highly volatile, keep the room well-ventilated. Cap the test tubes or bottles at times when the experiment is not being performed. The experiment should not be performed near any open flames.

8. Other properties, besides Δt values, vary with molecular size and consequent size of intermolecular forces of attraction. Viscosity increases noticeably from methanol through 1-butanol. The boiling temperatures of methanol, ethanol, 1-propanol, and 1-butanol are 65°C, 78°C, 97°C, and 117°C, respectively.

9. The stored calibration for the Stainless Steel Temperature Probe or Direct-Connect Temperature Probe works well for this experiment.

10. **HAZARD ALERTS:**

   n-Hexane: Flammable liquid: dangerous fire risk; may be irritating to respiratory tract. Hazard Code: B—Hazardous.

   Methanol: Flammable; dangerous fire risk; toxic by ingestion (ingestion may cause blindness). Hazard Code: B—Hazardous.

   Ethanol: Dangerous fire risk; flammable; addition of denaturant makes the product poisonous—it cannot be made non-poisonous; store in a dedicated flammables cabinet or safety cans. If a flammables cabinet or safety cans are not available, store in a Flinn *Saf-Stor®* Can. Hazard Code: C—Somewhat hazardous.

   n-Pentane: Flammable liquid; narcotic in high concentrations. Hazard Code: B—Hazardous.

   1-Propanol: Flammable liquid; dangerous fire risk; harmful to eyes and respiratory tract. Hazard Code: B—Hazardous.

   1-Butanol: Moderate fire risk; toxic on prolonged inhalation; eye irritant; absorbed by skin. Hazard Code: B—Hazardous.

   The hazard information reference is: Flinn Scientific, Inc., *Chemical & Biological Catalog Reference Manual*, 1-800-452-1261, www.flinnsci.com. See *Appendix D* of this book, *Chemistry with Vernier,* for more information.

11. One teacher has found that piping which can be purchased at a yard goods or sewing store serves as an appropriate sleeve for the temperature probe. You have to cut it pieces and remove the "rope" but then it works fine. It gives a nice consistent fit.

ANSWERS TO QUESTIONS

1. Even though n-pentane and 1-butanol have molecular weights of 72 and 74, respectively, 1-butanol has a much smaller Δt due to the presence of hydrogen bonding between its molecules. This results in a stronger attraction, and a slower rate of evaporation.

2. The 1-butanol has the strongest attractions between its molecules. Methanol has the weakest attractions. The 1-butanol has the largest molecules and resulting strongest dispersion forces. This gives it the lowest evaporation rate and the smallest Δt.

3. The n-hexane has the stronger attractions between its molecules. The n-pentane has the weaker attractions. The n-hexane has the larger molecules and the resulting stronger dispersion forces. This gives it a lower evaporation rate and the smallest Δt.

4. See the fourth sample graph on p. 9–4 T.

PRE-LAB RESULTS

Substance	Formula	Structural Formulas	Molecular Weight	Hydrogen Bond (Yes or No)
ethanol	C_2H_5OH	H H | | H–C–C–O–H | | H H	46	yes
1-propanol	C_3H_7OH	H H H | | | H–C–C–C–O–H | | | H H H	60	yes
1-butanol	C_4H_9OH	H H H H | | | | H–C–C–C–C–O–H | | | | H H H H	74	yes
n-pentane	C_5H_{12}	H H H H H | | | | | H–C–C–C–C–C–H | | | | | H H H H H	72	no
methanol	CH_3OH	H | H–C–O–H | H	32	yes
n-hexane	C_6H_{14}	H H H H H H | | | | | | H–C–C–C–C–C–C–H | | | | | | H H H H H H	86	no

DATA TABLE

Substance	t_1 (°C)	t_2 (°C)	Δt (t_1-t_2) (°C)
ethanol	23.5	15.2	8.3
1-propanol	23.0	18.1	4.9
1-butanol	23.2	21.5	1.7
n-pentane	23.0	6.9	16.1
Methanol	22.9	9.8	13.1
n-hexane	23.2	11.2	12.0

Predicted Δt (°C)	Explanation
varies (< 4.9°C)	It has a higher molecular wt. than 1-propanol (both have H-bonds).
varies (> 8.3°C)	It has a higher molecular wt. than either, but no H-bonding.
varies (> 8.3°C)	It has a lower molecular wt. than ethanol (both have H-bonds).
varies (< 16.1°C)	It has a higher molecular wt. than n-pentane; also no H-bonding.

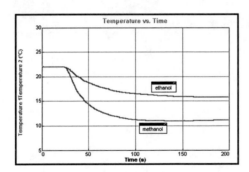

Evaporation of methanol and ethanol

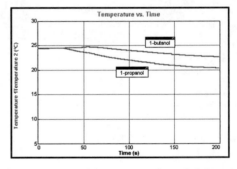

Evaporation of 1-propanol and 1-butanol

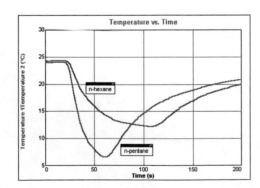

Evaporation of n-pentane and n-hexane

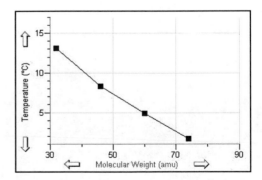

Temperature change vs. alcohol molecular wt.

Vapor Pressure of Liquids

In this experiment, you will investigate the relationship between the vapor pressure of a liquid and its temperature. When a liquid is added to the Erlenmeyer flask shown in Figure 1, it will evaporate into the air above it in the flask. Eventually, equilibrium is reached between the rate of evaporation and the rate of condensation. At this point, the vapor pressure of the liquid is equal to the partial pressure of its vapor in the flask. Pressure and temperature data will be collected using a Gas Pressure Sensor and a Temperature Probe. The flask will be placed in water baths of different temperatures to determine the effect of temperature on vapor pressure. You will also compare the vapor pressure of two different liquids, ethanol and methanol, at the same temperature.

OBJECTIVES

In this experiment, you will

- Investigate the relationship between the vapor pressure of a liquid and its temperature.
- Compare the vapor pressure of two different liquids at the same temperature.

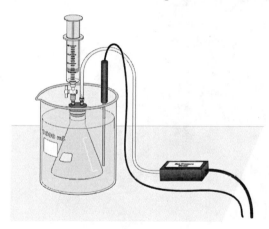

Figure 1

MATERIALS

computer	20 mL syringe
Vernier computer interface	two 125 mL Erlenmeyer flasks
Logger *Pro*	methanol
Vernier Gas Pressure Sensor	ethanol
Vernier Temperature Probe	ice
rubber-stopper assembly	four 1 liter beakers
plastic tubing with two connectors	

PROCEDURE

1. Obtain and wear goggles! **CAUTION:** The alcohols used in this experiment are flammable and poisonous. Avoid inhaling their vapors. Avoid contacting them with your skin or clothing. Be sure there are no open flames in the lab during this experiment. Notify your teacher immediately if an accident occurs.

2. Use 1 liter beakers to prepare four water baths, one in each of the following temperature ranges: 0 to 5°C, 10 to 15°C, 20 to 25°C (use room temperature water), and 30 to 35°C. For each water bath, mix varying amounts of warm water, cool water, and ice to obtain a volume of 800 mL in a 1 L beaker. To save time and beakers, several lab groups can use the same set of water baths.

3. Prepare the Temperature Probe and Gas Pressure Sensor for data collection.

   a. Plug the Gas Pressure Sensor into CH1 and the Temperature Probe into CH2 of the computer interface.

   b. Obtain a rubber-stopper assembly with a piece of heavy-wall plastic tubing connected to one of its two valves. Attach the connector at the free end of the plastic tubing to the open stem of the Gas Pressure Sensor with a clockwise turn. Leave its two-way valve on the rubber stopper open (lined up with the valve stem as shown in Figure 2) until Step 9.

   c. Insert the rubber-stopper assembly into a 125 mL Erlenmeyer flask. **Important:** Twist the stopper into the neck of the flask to ensure a tight fit.

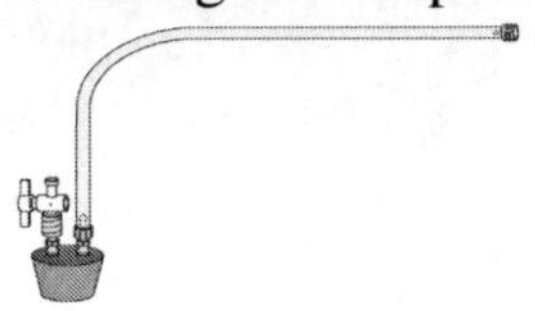

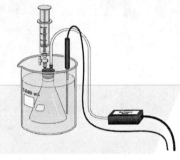

Figure 2

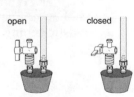

Figure 3

4. Prepare the computer for data collection by opening the file "10 Vapor Pressure" from the *Chemistry with Vernier* folder of Logger*Pro*.

5. The temperature and pressure readings should now be displayed in the meter. While the two-way valve above the rubber stopper is still open, record the value for atmospheric pressure in your data table (round to the nearest 0.1 kPa).

6. Finish setting up the apparatus shown in Figure 3:

   a. Obtain a room-temperature water bath (20–25°C).

   b. Place the Temperature Probe in the water bath.

   c. Hold the flask in the water bath, with the entire flask covered as shown in Figure 3.

   d. After 30 seconds, close the 2-way valve *above the rubber stopper* as shown in Figure 4—do this by turning the white valve handle so it is perpendicular with the valve stem itself.

Figure 4

7. Obtain the methanol container and the syringe. Draw 3 mL of the methanol up into the syringe. With the two-way valve still closed, screw the syringe onto the two-way valve, as shown in Figure 3.

8. Introduce the methanol into the Erlenmeyer flask.

    a. Open the 2-way valve above the rubber stopper—do this by turning the white valve handle so it is aligned with the valve stem (see Figure 4).
    b. Squirt the methanol into the flask by pushing in the plunger of the syringe.
    c. *Quickly* return the plunger of the syringe back to the 3 mL mark of the syringe, then close the 2-way valve by turning the white valve handle so it is perpendicular with the valve stem.
    d. Remove the syringe from the 2-way valve with a counter-clockwise turn.

9. To monitor and collect pressure and temperature data:

    a. Click ▶ Collect.
    b. When the pressure and temperature readings displayed in the meter stabilize, equilibrium between methanol liquid and vapor has been established. Click ⊕ Keep. The first pressure-temperature data pair is now stored.

10. To collect another data pair using the 30–35°C water bath:

    a. Place the Erlenmeyer flask assembly and the temperature probe into the 30–35°C water bath. Make sure the entire flask is covered.
    b. When the pressure and temperature readings displayed on the computer monitor stabilize, click ⊕ Keep. The second data pair has now been stored.

11. For Trial 3, repeat the Step-10 procedure, using the 10–15°C water bath. Then repeat the Step-10 procedure for Trial 4, using the 0–5°C water bath.

12. Click ■ Stop to end data collection. Record the pressure and temperature values in your data table, or, if directed by your instructor, print a copy of the table.

13. Gently loosen and remove the Gas Pressure Sensor so the Erlenmeyer flask is open to the atmosphere. Remove the stopper assembly from the flask and dispose of the methanol as directed by your teacher.

14. Obtain another clean, dry 125 mL Erlenmeyer flask. Draw air in and out of the syringe enough times that you are certain that all of the methanol has evaporated from it.

16. Repeat Steps 6–8 to do *one* trial only using ethanol in the room temperature water bath. When the pressure stabilizes, record the measured pressure of ethanol displayed in the meter in your data table.

17. Gently loosen and remove the stopper assembly from the flask and dispose of the ethanol as directed by your teacher.

PROCESSING THE DATA

1. Convert each of the Celsius temperatures to Kelvin (K). Write the answer in the space provided.

2. To obtain the vapor pressure of methanol and ethanol, the air pressure must be subtracted from each of the measured pressure values. However, for Trials 2–4, even if *no* methanol was present, the pressure in the flask would have increased due to a higher temperature, or decreased due to a lower temperature (remember those gas laws?). Therefore, you must convert the atmospheric pressure at the temperature of the *first* water bath to a *corrected* air pressure at the temperature of the water bath in Trial 2, 3, or 4. To do this, use the gas-law equation (use the Kelvin temperatures):

$$\frac{P_2}{T_2} = \frac{P_1}{T_1}$$

where P_1 and T_1 are the atmospheric pressure and the temperature of the Trial 1 (room temperature) water bath. T_2 is the temperature of the water bath in Trial 2, 3, or 4. Solve for P_2, and record this value as the *corrected* air pressure for Trials 2, 3, and 4. For Trial 1 of methanol and Trial 1 of ethanol, it is not necessary to make a correction; for these two trials, simply record the atmospheric pressure value in the blank designated for air pressure.

3. Obtain the vapor pressure by subtracting the corrected air pressure from the measured pressure in Trials 2-4. Subtract the uncorrected air pressure in Trial 1 of methanol (and Trial 1 of ethanol) from the measured pressure.

4. Plot a graph of vapor pressure *vs.* temperature (°C) for the four data pairs you collected for methanol. Temperature is the independent variable and vapor pressure is the dependent variable. As directed by your teacher, plot the graph manually, or use Logger *Pro*. Note: Be sure to plot the *vapor pressure*, not the *measured pressure*.

5. How would you describe the relationship between vapor pressure and temperature, as represented in the graph you made in the previous step? Explain this relationship using the concept of kinetic energy of molecules.

6. Which liquid, methanol or ethanol, had the larger vapor pressure value at room temperature? Explain your answer. Take into account various intermolecular forces in these two liquids.

DATA AND CALCULATIONS

Atmospheric pressure	_____ kPa

Substance	Methanol				Ethanol
Trial	1	2	3	4	1
Temperature (°C)	°C	°C	°C	°C	°C
Temperature (K)	K	K	K	K	K
Measured pressure	kPa	kPa	kPa	kPa	kPa

Air pressure	no correction	corrected	corrected	corrected	no correction
	kPa	kPa	kPa	kPa	kPa
Vapor pressure	kPa	kPa	kPa	kPa	kPa

EXTENSION

The Clausius-Clapeyron equation describes the relationship between vapor pressure and absolute temperature:

$$\ln P = \Delta H_{vap} / RT + B$$

where ln P is the natural logarithm of the vapor pressure, ΔH_{vap} is the heat of vaporization, T is the absolute temperature, and B is a positive constant. If this equation is rearranged in slope-intercept form ($y = mx + b$):

$$\ln P = \frac{\Delta H_{vap}}{R} \bullet \frac{1}{T} + B$$

the slope, m, should be equal to $-\Delta H_{vap}/R$. If a plot of ln P vs. $1/T$ is made, the heat of vaporization can be determined from the slope of the curve. Plot the graph using Logger *Pro*:

1. Go to Page 2 of the experiment file by clicking on the Next Page button, ⮕.

2. In the table, enter the four vapor pressure-temperature data pairs. To do this:

   a. Click on the first cell in the Temperature (K) data column in the table. Type in temperature value (K) for the first data pair, and press the ENTER key.

   b. The cursor will now be in the Vapor Pressure (kPa) data column—type in its value and press ENTER.

   c. Continue in this manner to enter the last three data pairs values.

   d. If necessary, click on the Autoscale button, 🅰, to automatically rescale the data points.

3. Create a column 1/Temperature (in units of 1/K). To do this:

   a. Choose New Calculated Column from the Data menu.

   b. Enter "1/Temperature" as the Name, "1/Temp" as the Short Name, and "1/K" as the Unit.

   c. Enter the correct formula for the column (1/Temperature) into the Equation edit box. To do this, type in "1" and "/". Then select "Temperature Kelvin" from the Variables list. In the Equation edit box, you should now see displayed: 1/"Temperature".

   d. Click ⬚ Done ⬚.

4. Create a column ln Vapor Pressure. To do this:

   a. Choose New Calculated Column from the Data menu.

   b. Enter "ln Vapor Pressure" as the Name, "ln V Press" as the Short Name. You do not need to enter a unit.

   c. Enter the correct formula for the column into the Equation edit box. Choose "ln" from the Function list. Then select "Vapor Pressure" from the Variables list. In the Equation edit box, you should now see displayed: ln("Vapor Pressure"). Click ⬚ Done ⬚.

   d. Click on the vertical axis label and choose ln Vapor Pressure.

   e. Click on the horizontal axis label and choose 1/Temperature.

   f. Autoscale the graph by clicking on the Autoscale button, 🅐, on the toolbar.

   g. Click the Linear Fit button, 📈.

5. From the Regression Statistics option, find the slope, *m*, of the regression line.

6. Use the slope value to calculate the heat of vaporization for methanol ($m = -\Delta H_{vap} / R$).

TEACHER INFORMATION

Vapor Pressure of Liquids

1. The student pages with complete instructions for data-collection using LabQuest App, Logger *Pro* (computers), EasyData or DataMate (calculators), and DataPro (Palm handhelds) can be found on the CD that accompanies this book. See *Appendix A* for more information.

2. Have a good supply of room-temperature tap water available for students to use for the Trial-1 water bath. Room-temperature water helps prevent the water-bath temperature from changing between Trial 1 of methanol and Trial 1 of ethanol. This improves the validity of comparing their vapor pressures.

3. Emphasize to your students the importance of providing an airtight fit with all plastic-tubing connections and when closing valves or twisting the stopper into the Erlenmeyer flask.

4. All of the pressure valves, tubing, and connectors used in this experiment are included with Vernier Pressure Sensors shipped after 1997. These accessories are also helpful when performing vapor pressure experiments such as Experiment 7 in this manual, *Pressure-Temperature Relationship in Gases*

   If you purchased your pressure sensor at an earlier date, or need replacement parts, Vernier has a Pressure Sensor Accessories Kit (PS-ACC) that includes all of the parts shown here for doing pressure-temperature experiments (except the Erlenmeyer flask). Using this kit allows for easy assembly of a completely airtight system. The kit includes the following parts:

   • two ribbed, tapered valve connectors inserted into a No. 5 rubber stopper
   • one ribbed, tapered valve connectors inserted into a No. 1 rubber stopper
   • two Luer-lock connectors connected to either end of a piece of plastic tubing
   • one two-way valve
   • one 20 mL syringe
   • two tubing clamps for transpiration experiments

5. Connect the piece of plastic tubing to the shorter valve connector on the rubber stopper prior to the start of class (as shown here). Threaded Luer-lock connectors are attached to both ends of the plastic tubing. Connect one of these to the valve stem with a 1/2 clockwise turn.

6. Emphasize to your students the importance of having an airtight fit after closing the 2-way valve and when twisting the stopper into the Erlenmeyer flask.

7. Even though this procedure requires students to calculate changes in total gas pressure due to changes in temperature, we felt it was less complicated than connecting the container to a vacuum pump or aspirator. Your students may be surprised when the pressure actually drops below the original air pressure at low temperatures.

8. If your class periods are longer than 50 minutes, you may choose to continue measuring the vapor pressure of ethanol in the remaining 3 water baths and then plot the vapor pressure-temperature curve on the same graph as methanol.

9. The Vernier temperature calibration for the Stainless Steel Temperature Probe works well for this experiment.

   The stored pressure calibration for the Gas Pressure Sensor also works well for this experiment.

10. Discuss with your students the relationship between vapor pressure and the rates of evaporation and condensation. It is important for students to relate the resulting equilibrium to the stabilizing of the pressure they observe in the experiment. Remind them to look for the presence of some remaining liquid once equilibrium is reached, since there can be no equilibrium without some liquid being present. **Note:** There is little chance of all of the liquid evaporating; approximately 1–2 mL of liquid will be delivered, but only about 0.05 mL will actually evaporate at the highest temperature.

11. **HAZARD ALERTS:**

   Methanol: Flammable; dangerous fire risk; toxic by ingestion (ingestion may cause blindness). Hazard Code: B—Hazardous.

   Ethanol: Dangerous fire risk; flammable; addition of denaturant makes the product poisonous—it cannot be made non-poisonous; store in a dedicated flammables cabinet or safety cans. If a flammables cabinet or safety cans are not available, store in a Flinn *Saf-Stor*® Can. Hazard Code: C—Somewhat hazardous.

   The hazard information reference is: Flinn Scientific, Inc., *Chemical & Biological Catalog Reference Manual,* (800) 452-1261, www.flinnsci.com. See *Appendix D* of this book, *Chemistry with Vernier*, for more information.

ANSWERS TO QUESTIONS

5. The vapor pressure of methanol increased as temperature increased, although not linearly (the curve gets steeper as temperature increases). As temperature increases, more molecules have enough kinetic energy to overcome the intermolecular force of attraction between molecules in the liquid.

6. Methanol has the higher vapor pressure. Although both molecules have hydrogen-bonding capability due to the -OH functional group, methanol has a lower molecular weight (32 amu) than ethanol (46 amu). As a result, methanol has weaker dispersion forces between molecules, and evaporates more easily.

SAMPLE RESULTS

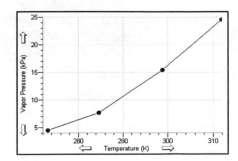

Vapor pressure (methanol) vs. temperature

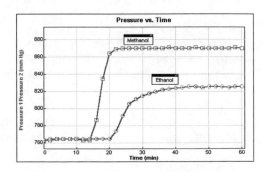

Measured pressure (total) vs. time

SAMPLE DATA AND CALCULATIONS

Atmospheric pressure	101.6 kPa

Substance	Methanol				Ethanol
Trial	1	2	3	4	1
Temperature (°C)	25.7°C	38.8°C	11.5°C	0.0°C	25.7°C
Temperature (K)	298.8 K	311.9 K	284.6 K	273.1 K	298.8 K
Measured pressure	117.0 kPa	130.6 kPa	104.5 kPa	97.0 kPa	110.0 kPa

Air pressure	no correction	corrected	corrected	corrected	no correction
		$\dfrac{(101.6)(312)}{(299)}$	$\dfrac{(101.6)(285)}{(299)}$	$\dfrac{(101.6)(273)}{(299)}$	
	101.6 kPa	106.0 kPa	96.8 kPa	92.5 kPa	101.6 kPa
Vapor pressure	117.0 – 101.6	130.6 – 106.0	104.5 – 96.8	97.0 – 92.5	110.0 – 101.6
	15.4 kPa	24.6 kPa	7.7 kPa	4.5 kPa	8.4 kPa

EXTENSION

Using the slope obtained from the graph of ln vapor pressure *vs.* 1/temperature (K) for methanol, the heat of vaporization can be calculated, in kJ/mol.

$$\Delta H_{vap} = -m \cdot R = -\,(-3779.8\ K)\,(8.31\ J/mol \cdot K) = 32330\ J/mol = 31.4\ kJ/mol$$

This value compares favorably with the accepted heat of vaporization of methanol, 34.4 kJ/mol.

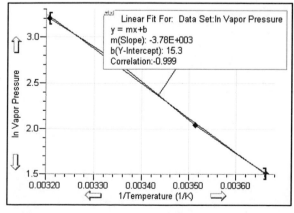

ln Vapor pressure vs. 1/Temperature (K)

Determining the Concentration of a Solution: Beer's Law

The primary objective of this experiment is to determine the concentration of an unknown nickel (II) sulfate solution. To accomplish this, you will use a Colorimeter or a Spectrometer to pass light through the solution, striking a detector on the opposite side. The wavelength of light used should be one that is absorbed by the solution. The $NiSO_4$ solution used in this experiment has a deep green color, so Colorimeter users will be instructed to use the red LED. Spectrometer users will determine an appropriate wavelength based on the absorbance spectrum of the solution. The light striking the detector is reported as *absorbance* or *percent transmittance*. A higher concentration of the colored solution absorbs more light (and transmits less) than a solution of lower concentration.

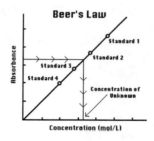

Figure 1

You are to prepare five nickel sulfate solutions of known concentration (standard solutions). Each is transferred to a small, rectangular cuvette that is placed into the Colorimeter or Spectrometer. The amount of light that penetrates the solution and strikes the detector is used to compute the absorbance of each solution. When a graph of absorbance *vs.* concentration is plotted for the standard solutions, a direct relationship should result, as shown in Figure 1. The direct relationship between absorbance and concentration for a solution is known as Beer's law.

The concentration of an *unknown* $NiSO_4$ solution is then determined by measuring its absorbance. By locating the absorbance of the unknown on the vertical axis of the graph, the corresponding concentration can be found on the horizontal axis (follow the arrows in Figure 1). The concentration of the unknown can also be found using the slope of the Beer's law curve.

OBJECTIVES

In this experiment, you will

- Prepare $NiSO_4$ standard solutions.
- Use a Colorimeter or Spectrometer to measure the absorbance value of each standard solution.
- Find the relationship between absorbance and concentration of a solution.
- Use the results of this experiment to determine the unknown concentration of another $NiSO_4$ solution.

MATERIALS

computer	30 mL of 0.40 M $NiSO_4$
Vernier computer interface*	5 mL of $NiSO_4$ unknown solution
Logger*Pro*	two 10 mL pipets (or graduated cylinders)
Colorimeter or Spectrometer	pipet pump or pipet bulb
one cuvette	distilled water
five 20 × 150 mm test tubes	test tube rack
tissues (preferably lint-free)	two 100 mL beakers
stirring rod	

* No interface is required if using a Spectrometer

PROCEDURE

Both Spectrometer and Colorimeter Users

1. Obtain and wear goggles! **CAUTION:** Be careful not to ingest any $NiSO_4$ solution or spill any on your skin. Inform your teacher immediately in the event of an accident.

2. Add about 30 mL of 0.40 M $NiSO_4$ stock solution to a 100 mL beaker. Add about 30 mL of distilled water to another 100 mL beaker.

3. Label four clean, dry, test tubes 1–4 (the fifth solution is the beaker of 0.40 M $NiSO_4$). Pipet 2, 4, 6, and 8 mL of 0.40 M $NiSO_4$ solution into Test Tubes 1–4, respectively. With a second pipet, deliver 8, 6, 4, and 2 mL of distilled water into Test Tubes 1–4, respectively. *Thoroughly* mix each solution with a stirring rod. Clean and dry the stirring rod between stirrings. Keep the remaining 0.40 M $NiSO_4$ in the 100 mL beaker to use in the fifth trial. Volumes and concentrations for the trials are summarized below:

Trial number	0.40 M $NiSO_4$ (mL)	Distilled H_2O (mL)	Concentration (M)
1	2	8	0.08
2	4	6	0.16
3	6	4	0.24
4	8	2	0.32
5	~10	0	0.40

4. Prepare a *blank* by filling a cuvette 3/4 full with distilled water. To correctly use cuvettes, remember:

   - Wipe the outside of each cuvette with a lint-free tissue.
   - Handle cuvettes only by the top edge of the ribbed sides.
   - Dislodge any bubbles by gently tapping the cuvette on a hard surface.
   - Always position the cuvette so the light passes through the clear sides.

Spectrometer Users Only (Colorimeter users proceed to the Colorimeter section)

5. Use a USB cable to connect the Spectrometer to the computer. Choose New from the File menu.

6. To calibrate the Spectrometer, place the blank cuvette into the cuvette slot of the Spectrometer, choose Calibrate ▶ Spectrometer from the Experiment menu. The calibration

dialog box will display the message: "Waiting 90 seconds for lamp to warm up." After 90 seconds, the message will change to "Warmup complete." Click 〔 OK 〕.

7. Determine the optimal wavelength for creating this standard curve.

   a. Remove the blank cuvette, and place the 0.40 M standard into the cuvette slot.

   b. Click 〔▸Collect〕. The absorbance *vs.* wavelength spectrum will be displayed. Click 〔■ Stop〕.

   c. To set up the data collection mode and select a wavelength for analysis, click on the Configure Spectrometer Data Collection icon, 🏠.

   d. Click Abs *vs.* Concentration (under the Set Collection Mode).

   e. Click 〔 Clear 〕 (Clear).

   f. Locate an absorbance peak in the red region of the spectrum and click on that peak value to select its wavelength. Click 〔 OK 〕.

   g. Proceed directly to Step 8.

Colorimeter Users Only

5. Connect the Colorimeter to the computer interface. Prepare the computer for data collection by opening the file "11 Beer's Law" from the *Chemistry with Vernier* folder of Logger*Pro*.

6. Open the Colorimeter lid, insert the blank, and close the lid.

7. To calibrate the Colorimeter, press the < or > button on the Colorimeter to select the wavelength of 635 nm (Red). Press the CAL button until the red LED begins to flash and then release the CAL button. When the LED stops flashing, the calibration is complete.

Both Spectrometer and Colorimeter Users

8. You are now ready to collect absorbance data for the five standard solutions. Click 〔▸Collect〕. Using the solution in Test Tube 1, rinse the cuvette twice with ~1 mL amounts and then fill it 3/4 full. Wipe the outside with a tissue and place it in the device (Colorimeter or Spectrometer). Close the lid on the Colorimeter. Wait for the absorbance value displayed on the monitor to stabilize. Then click 〔⊕ Keep〕 type **0.080** in the edit box, and press the ENTER key. The data pair you just collected will now be plotted on the graph.

9. Discard the cuvette contents as directed. Rinse the cuvette twice with the Test Tube 2 solution, 0.16 M NiSO$_4$, and fill the cuvette 3/4 full. Wipe the outside and place it in the device. When the absorbance value stabilizes, click 〔⊕ Keep〕, type **0.16** in the edit box, and press the ENTER key.

10. Repeat the Step 9 procedure to save and plot the absorbance and concentration values of the solutions in Test Tube 3 (0.24 M) and Test Tube 4 (0.32 M), as well as the stock 0.40 M NiSO$_4$. Wait until Step 13 to do the unknown. When you have finished with the 0.40 M NiSO$_4$ solution, click 〔■ Stop〕.

11. In your Data and Calculations table, record the absorbance and concentration data pairs that are displayed in the table.

12. Examine the graph of absorbance *vs.* concentration. To see if the curve represents a direct relationship between these two variables, click the Linear Fit button, 〔📈〕. A best-fit linear regression line will be shown for your five data points. This line should pass near or through the data points *and* the origin of the graph. (**Note:** Another option is to choose Curve Fit from the Analyze menu, and then select Proportional. The Proportional fit has a y-intercept

value equal to 0; therefore, this regression line will always pass through the origin of the graph).

13. Obtain about 5 mL of the unknown $NiSO_4$ in another clean, dry, test tube. Record the number of the unknown in the Data and Calculations table. Rinse the cuvette twice with the unknown solution and fill it about 3/4 full. Wipe the outside of the cuvette and place it in the device. Read the absorbance value displayed in the meter. (**Important:** The reading in the meter is live, so it is **not** necessary to click ▶ Collect to read the absorbance value.) When the displayed absorbance value stabilizes, record its value in Trial 6 of the Data and Calculations table.

14. Discard the solutions as directed by your teacher. Proceed directly to Steps 1 and 2 of Processing the Data.

PROCESSING THE DATA

1. Use the following method to determine the unknown concentration. With the linear regression curve still displayed on your graph, choose Interpolate from the Analyze menu. A vertical cursor now appears on the graph. The cursor's concentration and absorbance coordinates are displayed in the floating box. Move the cursor along the regression line until the absorbance value is approximately the same as the absorbance value you recorded in Step 13. The corresponding concentration value is the concentration of the unknown solution, in mol/L.

2. Print a graph of absorbance *vs.* concentration, with a regression line and interpolated unknown concentration displayed. To keep the interpolated concentration value displayed, move the cursor straight up the vertical cursor line until the tool bar is reached. Enter your name(s) and the number of copies of the graph you want.

DATA AND CALCULATIONS

Trial	Concentration (mol/L)	Absorbance
1	0.080	
2	0.16	
3	0.24	
4	0.32	
5	0.40	
6	Unknown number ____	

Concentration of unknown	mol/L

Determining the Concentration of a Solution: Beer's Law

1. The student pages with complete instructions for data-collection using LabQuest App, Logger *Pro* (computers), EasyData or DataMate (calculators), and DataPro (Palm handhelds) can be found on the CD that accompanies this book. See *Appendix A* for more information.

2. The light source for the nickel (II) sulfate solution is the red LED (635 nm). Since the $NiSO_4$ is green in color, the nearly monochromatic red light is readily absorbed by the solution.

3. The 0.40 M $NiSO_4$ solution can be prepared by using 10.51 g of $NiSO_4 \cdot 6H_2O$ per 100 mL. **HAZARD ALERT:** Toxic; avoid dispersing this substance; dispense with care; Nickel dust is a *possible carcinogen*. Hazard Code: B—Hazardous.

   The hazard information reference is: Flinn Scientific, Inc., *Chemical & Biological Catalog Reference Manual,* (800) 452-1261, www.flinnsci.com. See *Appendix D* of this book, *Chemistry with Vernier*, for more information.

4. Solutions of $Ni(NO_3)_2$ also work well, and can be prepared by using 11.63 g of solid $Ni(NO_3)_2 \cdot 6H_2O$ per 100 mL of solution.

5. Unknowns can be prepared by doing dilutions starting with the 0.40 M $NiSO_4$ stock solution. For example, to prepare a 0.22 M unknown, use 55 mL of the standard plus 45 mL of water:
$$(55 \text{ mL} / 100 \text{ mL})(.40 \text{ M}) = 0.22 \text{ M}$$

6. This experiment works well using solutions of green food coloring. A solution with an absorbance similar to 0.40 M $NiSO_4$ can be prepared by dissolving 8–9 drops of green Schilling Food Coloring in 1 liter of water. Check to see that the absorbance of this stock solution falls in the range of 0.40 to 0.80. Assign this solution a concentration of 100%. Students will follow the same procedure to dilute the stock solution to 80%, 60%, 40%, and 20%. If you use this method, have your students load the Exp 11b Colorimeter in the Experiment 11 folder in *Chemistry with Vernier*. This file has concentration scaled from 0 to 100% on the horizontal axis.

7. The cuvette must be from 55% to 100% full in order to get a valid absorbance reading. If students fill the cuvette 3/4 full, as described in the procedure, they should easily be in this range. To avoid spilling solution into the cuvette slot, remind students not to fill the cuvette
8. Since there is some variation in the amount of light absorbed by the cuvette if it is rotated 180°, you should use a water-proof marker to make a reference mark on the top edge of one of the clear sides of all cuvettes. Students are reminded in the procedure to align this mark with the white reference mark at the top of the cuvette slot on the Colorimeter.

8. The use of a single cuvette in the procedure is to eliminate errors introduced by slight variations in the absorbance of different plastic cuvettes. If one cuvette is used throughout the experiment by a student group, this variable is eliminated. The two rinses done prior to adding a new solution can be accomplished very quickly.

9. There are two models of Vernier Colorimeters. The first model (rectangular shape) has three wavelength settings, and the newest model (a rounded shape) has four wavelength settings. The 635 nm wavelength of either model is used in this experiment. The newer model is an auto-ID sensor and supports automatic calibration (pressing the CAL button on the

Colorimeter with a blank cuvette in the slot). If you have an older model Colorimeter, see www.vernier.com/til/1665.html for calibration information.

10. This experiment gives you a good opportunity to discuss the relationship between percent transmittance and absorbance. At the end of the experiment, students can click the Absorbance vertical-axis label of the graph, and choose Transmittance. The graph should now be transmittance *vs.* concentration. You can also discuss the mathematical relationship between absorbance and percent transmittance, as represented by either of these formulas:

$$A = \log(100 / \%T) \text{ or } A = 2 - \log\%T$$

SAMPLE RESULTS

Trial	Concentration (mol / L)	Absorbance
1	0.080	0.089
2	0.16	0.186
3	0.24	0.281
4	0.32	0.374
5	0.40	0.463
6	Unknown number 1	0.308

Concentration of the unknown	0.265 mol/L

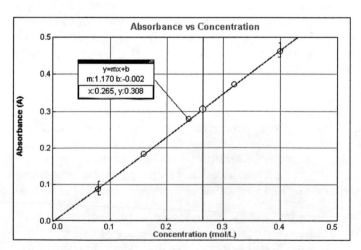

Absorbance vs. concentration for NiSO₄
with interpolation of the unknown displayed

Effect of Temperature on Solubility of a Salt

In this experiment, you will study the effect of changing temperature on the amount of solute that will dissolve in a given amount of water. Water solubility is an important physical property in chemistry, and is often expressed as the mass of solute that dissolves in 100 g of water at a certain temperature. In this experiment, you will completely dissolve different quantities of potassium nitrate, KNO_3, in the same volume of water at a high temperature. As each solution cools, you will monitor temperature using a computer-interfaced Temperature Probe and observe the precise instant that solid crystals start to form. At this moment, the solution is saturated and contains the maximum amount of solute at that temperature. Thus each data pair consists of a *solubility* value (g of solute per 100 g H_2O) and a corresponding *temperature*. A graph of the temperature-solubility data, known as a solubility curve, will be plotted using the computer.

OBJECTIVES

In this experiment, you will

- Study the effect of changing temperature on the amount of solute that will dissolve in a given amount of water.
- Plot a solubility curve.

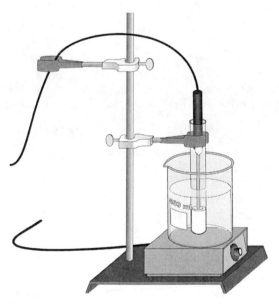

Figure 1

MATERIALS

computer	hot plate
Vernier computer interface	stirring rod
Logger *Pro*	potassium nitrate, KNO_3
Temperature Probe	distilled water
2 utility clamps	400 mL beaker
four 20×150 mm-test tubes	10 mL graduated cylinder or pipet
test tube rack	250 mL beaker
ring stand	

PROCEDURE

1. Obtain and wear goggles.

2. Label four test tubes 1-4. Into each of these test tubes, measure out the amounts of solid shown in the second column below (amount per 5 mL). Note: The third column (amount per 100 g of H_2O) is *proportional* to your measured quantity, and is the amount you will enter for your graph in Step 7 below.

Test tube number	Amount of KNO_3 used per 5 mL H_2O (weigh in Step 2)	Amount of KNO_3 used per 100 g H_2O (use in Step 10)
1	2.0	40
2	4.0	80
3	6.0	120
4	8.0	160

3. Add exactly 5.0 mL of distilled water to each test tube (assume 1.0 g/mL for water).

4. Connect the probe to the computer interface. Prepare the computer for data collection by opening the file "12 Temp and Solubility" from the *Chemistry with Vernier* folder of Logger *Pro*.

5. Fill a 400 mL beaker three-fourths full of tap water. Place it on a hot plate situated on (or next to) the base of a ring stand. Heat the water bath to about 90°C and adjust the heat to maintain the water at this temperature. Place the Temperature Probe in the water bath to monitor the temperature and to warm the probe. **CAUTION:** *To keep from damaging the Temperature Probe wire, hang it over another utility clamp pointing away from the hot plate, as shown in Figure 1.*

6. Use a utility clamp to fasten one of the test tubes to the ring stand. Lower the test tube into the water as shown in Figure 1. **Note:** In order to dissolve all of the KNO_3, Test Tubes 3 and 4 need to be heated to a higher temperature than Test Tubes 1 and 2. Use your stirring rod to stir the mixture until the KNO_3 is *completely* dissolved. Do not leave the test tube in the water bath any longer than is necessary to dissolve the solid.

7. When the KNO_3 is completely dissolved, click $\boxed{\blacktriangleright \text{Collect}}$. Remove the Temperature Probe from the water bath, wipe it dry, and place it into the solution in the test tube. Unfasten the utility clamp and test tube from the ring stand. Use the clamp to hold the test tube up to the light to look for the first sign of crystal formation. At the same time, stir the solution with a slight up and down motion of the Temperature Probe. At the moment crystallization starts to occur, click $\boxed{\circledS \text{ Keep}}$. Type the solubility value in the edit box (column 3 above, *g per 100 g H_2O*) and

press the ENTER key. Note: If you click ⊛ Keep too soon, you can press the ESC key to cancel the save. After you have saved the data pair, return the test tube to the test tube rack and place the Temperature Probe in the water bath for the next trial.

8. Repeat Steps 6 and 7 for each of the other three test tubes. Here are some suggestions to save time:

   - One lab partner can be stirring the next KNO_3-water mixture until it dissolves while the other partner watches for crystallization and enters data pairs using the computer.
   - Test Tubes 1 and 2 may be cooled to lower temperatures using cool tap water in the 250 mL beaker. This drops the temperature much faster than air. If the crystals form too quickly, *briefly* warm the test tube in the hot-water bath and redissolve the solid. Then repeat the cooling and collect the data pair.

9. When you have finished collecting data, click ■ Stop . Record the temperature (in °C) from the four trials in your data table, or, if directed by your instructor, print a copy of the table. Discard the four solutions as directed by your teacher.

10. Print a copy of the graph. Enter your name(s) and the number of copies you want to print.

PROCESSING THE DATA

1. Draw a best-fit curve for your data points on the printed graph.

2. According to your data, how is solubility of KNO_3 affected by an increase in temperature of the solvent?

3. Using your printed graph, tell if each of these solutions would be saturated or unsaturated:

   a. 110 g of KNO_3 in 100 g of water at 40°C
   b. 60 g of KNO_3 in 100 g of water at 70°C
   c. 140 g of KNO_3 in 200 g of water at 60°C

4. According to your graph, will 50 g of KNO_3 completely dissolve in 100 g of water at 50°C? Explain.

5. According to your graph, will 120 g of KNO_3 completely dissolve in 100 g of water at 40°C? Explain.

6. According to your graph, about how many grams of KNO_3 will dissolve in 100 g of water at 30°C?

DATA TABLE

Trial	Solubility (g / 100 g H_2O)	Temp (°C)
1	40.0	
2	80.0	
3	120.0	
4	160.0	

TEACHER INFORMATION

Effect of Temperature on Solubility of a Salt

1. The student pages with complete instructions for data-collection using LabQuest App, Logger *Pro* (computers), EasyData or DataMate (calculators), and DataPro (Palm handhelds) can be found on the CD that accompanies this book. See *Appendix A* for more information.

2. Have 20 g of solid KNO_3 available per student lab team. **HAZARD ALERT:** Potassium nitrate is a strong oxidant; fire and explosion risk when heated or in contact with organic material; skin irritant; highly toxic. Hazard Code: B—Hazardous.

   The hazard information reference is: Flinn Scientific, Inc., *Chemical & Biological Catalog Reference Manual,* (800) 452-1261, www.flinnsci.com. See *Appendix D* of this book, *Chemistry with Vernier*, for more information.

2. Students should be well-prepared for this experiment in order to comfortably complete it during a 50 minute class period. Remind them of the time-saving tips in the student procedure. Starting hot-water baths promptly will save time. The 2 g and 4 g samples do not need to be heated above 60°C in order to have all of the solid dissolve; thus, the water bath does not need to be at 90°C for these two trials. Toward the end of the class period, some students may have to use cool-water baths, especially for the 2 g and 4 g samples.

3. When using hot plates, caution students to keep the temperature probe wire well away from the hot plate.

4. The stored calibrations for the Stainless Steel Temperature Probe works well for this experiment.

ANSWERS TO QUESTIONS

2. Solubility of KNO_3 increases with increasing temperature.

3. Saturated

   Unsaturated

   Unsaturated

4. Yes, it will dissolve, since this point is below the solubility curve.

5. No, it will not dissolve, since this point is above the solubility curve.

6. At 30°C, ~50 g of KNO_3 would dissolve in 100 g of water.

SAMPLE DATA

Trial	Solubility (g / 100 g H_2O)	Temp (°C)
1	40.0	25.5
2	80.0	47.2
3	120.0	64.0
4	160.0	76.4

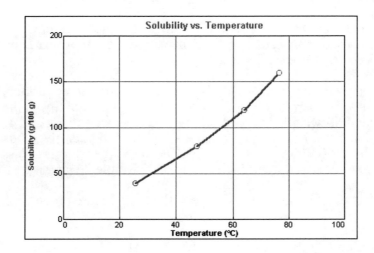

Properties of Solutions: Electrolytes and Non-Electrolytes

In this experiment, you will discover some properties of strong electrolytes, weak electrolytes, and non-electrolytes by observing the behavior of these substances in aqueous solutions. You will determine these properties using a Conductivity Probe. When the probe is placed in a solution that contains ions, and thus has the ability to conduct electricity, an electrical circuit is completed across the electrodes that are located on either side of the hole near the bottom of the probe body (see Figure 1). This results in a conductivity value that can be read by the computer. The unit of conductivity used in this experiment is the microsiemens per centimeter, or μS/cm.

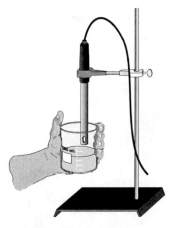

Figure 1

The size of the conductivity value depends on the ability of the aqueous solution to conduct electricity. Strong electrolytes produce large numbers of ions, which results in high conductivity values. Weak electrolytes result in low conductivity, and non-electrolytes should result in no conductivity. In this experiment, you will observe several factors that determine whether or not a solution conducts, and if so, the relative magnitude of the conductivity. Thus, this simple experiment allows you to learn a great deal about different compounds and their resulting solutions.

In each part of the experiment, you will be observing a different property of electrolytes. Keep in mind that you will be encountering three types of compounds and aqueous solutions:

Ionic Compounds

These are usually strong electrolytes and can be expected to 100% dissociate in aqueous solution.

> Example: $NaNO_3(s) \longrightarrow Na^+(aq) + NO_3^-(aq)$

Molecular Compounds

These are usually non-electrolytes. They do not dissociate to form ions. Resulting solutions do not conduct electricity.

> Example: $CH_3OH(l) \longrightarrow CH_3OH(aq)$

Molecular Acids

These are molecules that can partially or wholly dissociate, depending on their strength.

Example: Strong electrolyte $H_2SO_4 \longrightarrow H^+(aq) + HSO_4^-(aq)$ (100% dissociation)

Example: Weak electrolyte $HF \longleftrightarrow H^+(aq) + F^-(aq)$ (<100% dissociation)

OBJECTIVES

In this experiment, you will

- Write equations for the dissociation of compounds in water.
- Use a Conductivity Probe to measure the conductivity of solutions.
- Determine which molecules or ions are responsible for conductivity of solutions.
- Investigate the conductivity of solutions resulting from compounds that dissociate to produce different numbers of ions.

MATERIALS

computer	H_2O (distilled)
Vernier computer interface	0.05 M NaCl
Logger Pro	0.05 M $CaCl_2$
Vernier Conductivity Probe	0.05 M $AlCl_3$
250 mL beaker	0.05 M $HC_2H_3O_2$
wash bottle with distilled water	0.05 M H_3PO_4
tissues	0.05 M H_3BO_3
ring stand	0.05 M HCl
utility clamp	0.05 M CH_3OH (methanol)
H_2O (tap)	

PROCEDURE

1. Obtain and wear goggles! **CAUTION:** Handle the solutions in this experiment with care. Do not allow them to contact your skin. Notify your teacher in the event of an accident.

2. The Conductivity Probe is already attached to the interface. It should be set on the 0–20000 µS/cm position.

3. Prepare the computer to monitor conductivity by opening the file "13 Electrolytes" from the *Chemistry with Vernier* folder.

4. Obtain the Group A solution containers. The solutions are: 0.05 M NaCl, 0.05 M $CaCl_2$, and 0.05 M $AlCl_3$.

5. Measure the conductivity for each of the solutions.

   a. Carefully raise each vial and its contents up around the Conductivity Probe until the hole near the probe end is completely submerged. **Important:** Since the two electrodes are positioned on either side of the hole, this part of the probe must be completely submerged.

   b. Briefly swirl the beaker contents. When the reading has stabilized, record the value.

   c. Before testing the next solution, clean the electrodes by surrounding them with a 250 mL beaker and rinsing them with distilled water. Blot the outside of the probe end dry using a tissue. It is *not* necessary to dry the *inside* of the hole near the probe end.

6. Obtain the four Group B solution containers. These include 0.05 M H_3PO_4, 0.05 M $HC_2H_3O_2$, 0.05 M H_3BO_3, and 0.05 M HCl. Repeat the Step 5 procedure.

7. Obtain the five Group C solutions or liquids. These include 0.05 M CH_3OH, 0.05 M $C_2H_6O_2$, distilled H_2O, and tap H_2O. Repeat the Step 5 procedure.

DATA TABLE

Solution	Conductivity (µS/cm)
A - $CaCl_2$	
A - $AlCl_3$	
A - NaCl	
B - $HC_2H_3O_2$	
B - HCl	
B - H_3PO_4	
B - H_3BO_3	
C - $H_2O_{distilled}$	
C - H_2O_{tap}	
C - CH_3OH	

PROCESSING THE DATA

1. Based on your conductivity values, do the Group A compounds appear to be molecular, ionic, or molecular acids? Would you expect them to partially dissociate, completely dissociate, or not dissociate at all?

2. Why do the Group A compounds, each with the same concentration (0.05 M), have such large differences in conductivity values? **Hint:** Write an equation for the dissociation of each. Explain.

3. In Group B, do all four compounds appear to be molecular, ionic, or molecular acids? Classify each as a strong or weak electrolyte, and arrange them from the strongest to the weakest, based on conductivity values.

4. Write an equation for the dissociation of each of the compounds in Group B. Use $\longrightarrow$ for strong; $\longleftrightarrow$ for weak.

5. For H_3PO_4 and H_3BO_3, does the subscript "3" of hydrogen in these two formulas seem to result in additional ions in solution as it did in Group A? Explain.

6. In Group C, do all four compounds appear to be molecular, ionic, or molecular acids? Based on this answer, would you expect them to dissociate?

7. How do you explain the relatively high conductivity of tap water compared to a low or zero conductivity for distilled water?

Properties of Solutions: Electrolytes and Non-Electrolytes

1. The student pages with complete instructions for data-collection using LabQuest App, Logger *Pro* (computers), EasyData or DataMate (calculators), and DataPro (Palm handhelds) can be found on the CD that accompanies this book. See *Appendix A* for more information.

2. We suggest that you set up the Conductivity Probes before the experiment. Set the selection switch on the amplifier box of the probe to the 0–20000 μS/cm range.

3. Fewer sets of Groups A, B, and C can be prepared if students are advised that they need not start with Group A. Add solutions to 100 mL beakers or small vials to a depth that easily allows the hole near the Conductivity Probe tip to be completely submerged (the graphite electrodes of the probe are located on either side of this hole).

4. Preparation of solutions (prepare all solutions in distilled water):

    0.050 M $CaCl_2$ (5.55 g of solid calcium chloride, $CaCl_2$, per 1 L solution) Hazard Code: D—Relatively non-hazardous. Alternatively, 7.35 g $CaCl_2 \cdot 2H_2O$, per 1 L solution. **HAZARD ALERT:** Toxic by ingestion. Hazard Code: D—Relatively non-hazardous.

    0.050 M NaCl (2.93 g of solid sodium chloride, NaCl, per 1 L solution) **HAZARD ALERT:** Moderately toxic. Hazard Code: D—Relatively non-hazardous.

    0.050 M $AlCl_3$ (12.05 g of solid aluminum chloride, $AlCl_3 \cdot 6H_2O$, per 1 L solution)— preferred. Hazard Code: D—Relatively non-hazardous. Alternatively, 6.67 g anhydrous $AlCl_3$ per liter of solution. **HAZARD ALERT:** Reacts very violently with water; toxic by inhalation and ingestion; strong skin irritant. Hazard Code: A—Extremely hazardous.

    0.050 M HCl (4.2 mL of concentrated hydrochloric acid, HCl, per 1 L solution) **HAZARD ALERT:** Highly toxic by ingestion or inhalation; severely corrosive to skin and eyes. Hazard Code: A—Extremely hazardous.

    0.050 M $HC_2H_3O_2$ (2.9 mL of concentrated acetic acid, $HC_2H_3O_2$, per 1 L solution) **HAZARD ALERT:** Corrosive to skin and tissue; moderate fire risk (flash point: 39°C); moderately toxic by ingestion and inhalation. Hazard Code: A—Extremely hazardous.

    0.050 M H_3PO_4 (3.4 mL of concentrated phosphoric acid, H_3PO_4, per 1 L solution) **HAZARD ALERT:** Skin and eye irritant; moderately toxic by ingestion and inhalation; corrosive; burns tissue. Hazard Code: A—Extremely hazardous.

    0.050 M H_3BO_3 (3.09 g of solid boric acid, H_3BO_3, per 1 L solution) **HAZARD ALERT:** Moderately toxic by ingestion; irritant to skin in dry form. Hazard Code: C—Somewhat hazardous.

    0.050 M CH_3OH (1.60 g (2.1 mL) methanol per 1 L solution) **HAZARD ALERT:** Flammable; dangerous fire risk; toxic by ingestion (ingestion may cause blindness). Hazard Code: B—Hazardous.

    The hazard information reference is: Flinn Scientific, Inc., *Chemical & Biological Catalog Reference Manual*, (800) 452-1261, www.flinnsci.com. See *Appendix D of* this book, *Chemistry with Vernier*, for more information.

5. The procedure has students record conductivity values from the live readouts (without starting data collection). If you are using Logger *Pro* for data collection, another possibility is to have students use the Selected Events mode for each of the 11 trials. In Logger*Pro,* the file for this experiment is already set up for this option. Simply have your students click ▶ Collect , then click ⊕ Keep when the conductivity reading is stable. This saves the conductivity reading along with its trial number.

6. Conductivity readings are normally reported in microsiemens per centimeter, or μS/cm. This SI derived unit has replaced the conductivity unit, micromho/cm.

7. Students are instructed to rinse the probe with distilled water between samples. They are told to blot the probe tip dry—however, the directions also remind them that they do *not* need to blot dry the inside of the hole containing the graphite electrodes. It is cumbersome to do so, and leaving a drop or two of distilled water does not significantly dilute the next sample.

8. Using the stored calibration, measured conductivity values for H_3BO_3, CH_3OH, or distilled water will be in the range of 0 to 30 μS/cm. If a two-point calibration is performed, students will get readings closer to 0 μS/cm. These four samples will usually have a small conductivity value due to dissolved carbon dioxide, which forms aqueous ions according to the equation:

$$CO_2(g) + H_2O(l) \longleftrightarrow H^+(aq) + HCO_3^-(aq)$$

The resulting conductivity, usually about 1–3 μS/cm, can be accurately measured using the narrower 0–200 μS/cm setting and calibration for the Conductivity Probe. You could do this as a teacher demonstration, or instruct your students to do it as an extension to the experiment.

At the 0–200 μS/cm setting, students will also notice that the conductivity of boric acid is higher than distilled water, 0.05 M methanol, or 0.05 M ethylene glycol. This way, they can see that boric acid is a weak acid that ionizes to a very small extent. For example, we get a reading of 3.2 μS/cm for 0.05 M boric acid, but only 1.0 μS/cm for distilled water, and 1.0 μS/cm for 0.05 M methanol, using the 0–200 μS/cm setting.

9. If you wish to calibrate the Conductivity Probe to improve conductivity readings at low concentrations (as discussed in item 8 above), follow these directions:

First Calibration Point

a. Set up the data-collection software to calibrate the Conductivity Probe.

b. For the first calibration point, the Conductivity Probe should simply be in the air (out of any liquid or solution).

c. Type **0** in the edit box as the conductivity value (in μS/cm).

d. Wait until the voltage stabilizes, then Keep the point.

Second Calibration Point

e. Place the Conductivity Probe into a standard solution that is equivalent to 10,000 μS/cm. **Note:** This standard can be prepared by dissolving 5.566 g of solid sodium chloride, NaCl, in enough distilled water for 1 liter of solution.

f. Type **10000** in the edit box as the conductivity value for the second calibration point (in μS/cm).

g. Wait until the voltage stabilizes, then Keep the point. Then select either OK or Done depending on the software. This completes the calibration.

ANSWERS TO QUESTIONS

1. All three are ionic. They completely dissociate in water.

2. $AlCl_3 \longrightarrow Al^{3+} + 3\,Cl^-$ (4 moles of ions per mole)

    $CaCl_2 \longrightarrow Ca^{2+} + 2\,Cl^-$ (3 moles of ions per mole)

    $NaCl \longrightarrow Na^+ + Cl^-$ (2 moles of ions per mole)

    Even though all three solutions have the same initial concentration, 0.05 M, $AlCl_3$ dissociates to yield the largest number of moles of ions per mole, and as a result exhibits the highest conductivity in this series. $CaCl_2$ is next, and NaCl yields the fewest moles of ions per mole.

3. All three are molecular acids. HCl is a strong acid. H_3PO_4 is borderline between strong and weak, but is usually classified as a weak acid. Acetic acid, $HC_2H_3O_2$ is the next weakest acid and H_3BO_3 is the weakest.

4. $HCl \longrightarrow H^+ + Cl^-$ $H_3PO_4 \longleftrightarrow H^+ + H_2PO_4^-$

    $HC_2H_3O_2 \longleftrightarrow H^+ + C_2H_3O_2^-$ $H_3BO_3 \longleftrightarrow H^+ + H_2BO_3^-$

5. Since H_3PO_4 and H_3BO_3 are two of the weak acids in this series, one would conclude that the subscript "3" contributes little to their strengths. The equations for their dissociations indicate that only one H^+ dissociates to any appreciable extent from either of these weak acids. The dissociations of the second and third H^+ ions are insignificant by comparison.

6. All four compounds in Group C are molecular. None of them dissociates significantly.

7. Even though the water itself is molecular, it contains ionic impurities, such as Ca^{2+}, Mg^{2+}, HCO_3^-, and Cl^-. The ionic impurities contribute significantly to the conductivity of the solution. These ionic impurities have been removed from distilled water.

SAMPLE RESULTS

Solution	Conductivity (μS/cm)
A - $CaCl_2$	9362
A - NaCl	5214
A - $AlCl_3$	11707
B - $HC_2H_3O_2$	461
B - HCl	17330
B - H_3PO_4	6661
B - H_3BO_3	0
C - $H_2O_{distilled}$	0
C - H_2O_{tap}	(varies) 20 – 1000
C - CH_3OH	0

Conductivity of Solutions: The Effect of Concentration

If an ionic compound is dissolved in water, it dissociates into ions and the resulting solution will conduct electricity. Dissolving solid sodium chloride in water releases ions according to the equation:

$$NaCl(s) \longrightarrow Na^+(aq) + Cl^-(aq)$$

In this experiment, you will study the effect of increasing the concentration of an ionic compound on conductivity. Conductivity will be measured as concentration of the solution is gradually increased by the addition of concentrated NaCl drops. The same procedure will be used to investigate the effect of adding solutions with the same concentration (1.0 M), but different numbers of ions in their formulas: aluminum chloride, $AlCl_3$, and calcium chloride, $CaCl_2$. A computer-interfaced Conductivity Probe will be used to measure conductivity of the solution. Conductivity is measured in microsiemens per centimeter (μS/cm).

OBJECTIVES

In this experiment, you will

- Use a Conductivity Probe to measure the conductivity of solutions.
- Investigate the relationship between the conductivity and concentration of a solution.
- Investigate the conductivity of solutions resulting from compounds that dissociate to produce different number of ions.

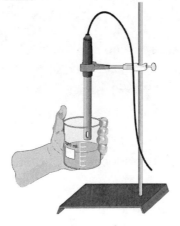

Figure 1

MATERIALS

computer	distilled water
Vernier computer interface	100 mL beaker
Logger *Pro*	1.0 M NaCl solution
Vernier Conductivity Probe	1.0 M $CaCl_2$ solution
ring stand	1.0 M $AlCl_3$ solution
utility clamp	stirring rod
wash bottle	tissue

PROCEDURE

1. Obtain and wear goggles.

2. Your experiment setup should look like Figure 1. The Conductivity Probe is already attached to the interface. It should be set on the 0–2000 µS/cm position..

3. Prepare the computer for data collection by opening the file "14 Conductivity Solutions" from the *Chemistry with Vernier* folder.

4. Add 70 mL of distilled water to a clean 100 mL beaker. Obtain a dropper bottle that contains 1.0 M NaCl solution.

5. Before adding any drops of solution:

   a. Click ▶ Collect .
   b. Carefully raise the beaker and its contents up around the Conductivity Probe until the hole near the probe end is completely submerged in the solution being tested. **Important:** Since the two electrodes are positioned on either side of the hole, this part of the probe must be completely submerged.
   c. Monitor the conductivity of the distilled water until the conductivity reading stabilizes.
   d. Click ⊛ Keep , and then lower the beaker away from the probe. Type **0** in the edit box (for 0 drops added). Press the ENTER key to store this data pair. This gives the conductivity of the water before any salt solution is added.

6. You are now ready to begin adding salt solution.

   a. Add 1 drop of NaCl solution to the distilled water. Stir to ensure thorough mixing.
   b. Raise the beaker until the hole near the probe end is completely submerged in the solution. Swirl the solution briefly.
   c. Monitor the conductivity of the solution until the reading stabilizes.
   d. Click ⊛ Keep , and then lower the beaker away from the probe. Type **1** (the total drops added) in the edit box and press ENTER.

7. Repeat the Step 6 procedure, entering **2** this time.

8. Continue this procedure, adding 1-drop portions of NaCl solution, measuring conductivity, and entering the total number of drops added—until a total of 8 drops have been added.

9. Click ■ Stop when you have finished collecting data. Dispose of the beaker contents as directed by your teacher. Rinse the probe tip with distilled water from a wash bottle. Carefully blot the probe dry with a tissue.

10. Prepare the computer for data collection. From the Experiment menu, choose Store Latest Run. This stores the data so it can be used later, but it will be still be displayed while you do your second and third trials.

11. Repeat Steps 4–10, this time using 1.0 M $AlCl_3$ solution in place of 1.0 M NaCl solution.

12. Repeat Steps 4–9, this time using 1.0 M $CaCl_2$ solution.

13. Click on the Linear Fit button, ⊠. Be sure all three data runs are selected, then click   OK . A best-fit linear regression line will be shown for each of your three runs. In your data table,

record the value of the slope, *m*, for each of the three solutions. (The linear regression statistics are displayed in a floating box for each of the data sets.)

14. To print a graph of concentration *vs.* volume showing all three data runs:

    a. Label all three curves by choosing Text Annotation from the Insert menu, and typing "sodium chloride" (or "aluminum chloride", or "calcium chloride") in the edit box. Then drag each box to a position near its respective curve.

    b. Print a copy of the graph, with all three data sets and the regression lines displayed. Enter your name(s) and the number of copies of the graph you want.

DATA TABLE

Solution	Slope, *m*
1.0 M NaCl	
1.0 M AlCl$_3$	
1.0 M CaCl$_2$	

PROCESSING THE DATA

1. Describe the appearance of each of the three curves on your graph.

2. Describe the change in conductivity as the concentration of the NaCl solution was increased by the addition of NaCl drops. What kind of mathematical relationship does there appear to be between conductivity and concentration?

3. Write a chemical equation for the dissociation of NaCl, AlCl$_3$, and CaCl$_2$ in water.

4. Which graph had the largest slope value? The smallest? Since all solutions had the same original concentration (1.0 M), what accounts for the difference in the slope of the three plots? Explain.

Conductivity of Solutions:
The Effect of Concentration

1. The student pages with complete instructions for data-collection using LabQuest App, Logger *Pro* (computers), EasyData or DataMate (calculators), and DataPro (Palm handhelds) can be found on the CD that accompanies this book. See *Appendix A* for more information.

2. We suggest that you set up the Conductivity Probes before the experiment. Set the selection switch on the amplifier box of the probe to the 0–2000 µS/cm range.

3. Distilled water and tissue can be used to clean the Conductivity Probe. See the Conductivity Probe booklet that comes with the Conductivity Probe for information on how the probes work, how to care for the probes, and calibrations.

4. All solutions are 1.0 M concentration. Have them available in dropper bottles (prepare all solutions in distilled water):

    1.0 M $CaCl_2$ (11.1 g of solid calcium chloride, $CaCl_2$, per 100 mL of solution) Hazard Code: D—Relatively non-hazardous. Alternatively, 14.7 g $CaCl_2 \cdot 2H_2O$, per 100 mL of solution. **HAZARD ALERT:** Toxic by ingestion. Hazard Code: D—Relatively non-hazardous.

    1.0 M NaCl (5.85 g of solid sodium chloride, NaCl, per 100 mL solution) **HAZARD ALERT:** Moderately toxic. Hazard Code: D—Relatively non-hazardous.

    1.0 M $AlCl_3$ (24.15 g of solid aluminum chloride, $AlCl_3 \cdot 6H_2O$, per 100 mL of solution)—preferred. Hazard Code: D—Relatively non-hazardous. Alternatively, 13.35 g anhydrous $AlCl_3$ per 100 mL of solution. **HAZARD ALERT:** Reacts very violently with water; toxic by inhalation and ingestion; strong skin irritant. Hazard Code: A—Extremely hazardous.

    The hazard information reference is: Flinn Scientific, Inc., *Chemical & Biological Catalog Reference Manual*, (800) 452-1261, www.flinnsci.com. See *Appendix D* of this book, *Chemistry with Vernier*, for more information.

5. For consistent results, students should dispense drops with the dropper bottle held in a vertical position.

6. Conductivity readings are normally reported in microsiemens per centimeter, or µS/cm. This SI derived unit has replaced the conductivity unit, micromho/cm.

7. Note that the ratio of slopes of NaCl, $CaCl_2$, and $AlCl_3$ is quite consistent with the ratio of ions produced upon dissociation:

    *   Ratio of slopes: 91.5 to 158.4 to 215.0
    *   Ratio of moles of ions, upon dissociation: 2 to 3 to 4

ANSWERS TO QUESTIONS

1. At low concentrations, each curve is nearly linear. The slope value was different for each of the three solutions: $AlCl_3$ was highest, $CaCl_2$ second highest, and NaCl lowest.

2. Conductivity increases as concentration is increased. The relationship appears to be direct.

3. $NaCl \longrightarrow Na^+ + Cl^-$ (2 moles of ions per mole)

   $AlCl_3 \longrightarrow Al^{3+} + 3\ Cl^-$ (4 moles of ions per mole)

   $CaCl_2 \longrightarrow Ca^{2+} + 2\ Cl^-$ (3 moles of ions per mole)

4. $AlCl_3$ has the largest slope value, NaCl the smallest. Even though all three solutions have the same initial concentration, 1.0 M, $AlCl_3$ dissociates to yield the largest number of moles of ions per mole (4). This results in $AlCl_3$ yielding more ions in solution, and the largest slope in this series. $CaCl_2$ is next with 3 moles of ions per mole, and NaCl yields the fewest, 2.

SAMPLE RESULTS

Solution	Slope, m
1.0 M NaCl	91.5
1.0 M AlCl$_3$	205.0
1.0 M CaCl$_2$	158.4

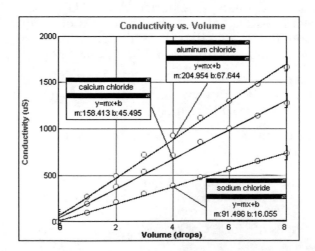

Using Freezing-Point Depression to Find Molecular Weight

When a solute is dissolved in a solvent, the freezing temperature is lowered in proportion to the number of moles of solute added. This property, known as freezing-point depression, is a *colligative property*; that is, it depends on the ratio of solute and solvent particles, not on the nature of the substance itself. The equation that shows this relationship is:

$$\Delta T = K_f \cdot m$$

where ΔT is the freezing point depression, K_f is the freezing point depression constant for a particular solvent (3.9°C-kg/mol for lauric acid in this experiment[1]), and m is the molality of the solution (in mol solute/kg solvent).

In this experiment, you will first find the freezing temperature of the pure solvent, lauric acid, $CH_3(CH_2)_{10}COOH$. You will then add a known mass of benzoic acid solute, C_6H_5COOH, to a known mass of lauric acid, and determine the lowering of the freezing temperature of the solution. In an earlier experiment, you observed the effect on the cooling behavior at the freezing point of adding a solute to a pure substance. By measuring the freezing point depression, ΔT, and the mass of benzoic acid, you can use the formula above to find the molecular weight of the benzoic acid solute, in g/mol.

OBJECTIVES

In this experiment, you will

- Determine the freezing temperature of pure lauric acid.
- Determine the freezing temperature of a solution of benzoic acid and lauric acid.
- Examine the freezing curves for each.
- Calculate the experimental molecular weight of benzoic acid.
- Compare it to the accepted molecular weight for benzoic acid.

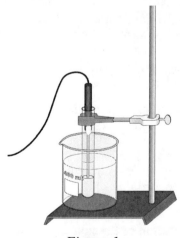

Figure 1

---

[1] "The Computer-Based Laboratory," *Journal of Chemical Education: Software*, 1988, Vol. 1A, No. 2, p. 73.

MATERIALS

computer
Vernier computer interface
Logger *Pro*
Temperature Probe
400 mL beaker
ring stand

utility clamp
18 × 150 mm test tube
lauric acid
benzoic acid
thermometer

PROCEDURE

1. Obtain and wear goggles.

2. Connect the Temperature Probe to the computer interface. Prepare the computer for data collection by opening the file "15 Freezing Pt Depression" from the *Chemistry with Vernier* folder.

Part I Freezing Temperature of Pure Lauric Acid

3. Add about 300 mL of tap water with a temperature between 20 and 25°C to a 400 mL beaker. Place the beaker on the base of the ring stand.

4. Use a utility clamp to obtain a test tube containing hot melted lauric acid from your teacher. Fasten the utility clamp at the top of the test tube. **CAUTION:** *Be careful not to spill the hot lauric acid on yourself and do not touch the bottom of the test tube.*

5. Insert the Temperature Probe into the hot lauric acid. About 30 seconds are required for the probe to warm up to the temperature of its surroundings and give correct temperature readings. During this time, fasten the utility clamp to the ring stand so the test tube is above the water bath. Then click ▶ Collect to begin data collection.

6. Lower the test tube into the water bath. Make sure the water level outside the test tube is higher than the lauric acid level inside the test tube. If the lauric acid is not above 50°C, obtain another lauric acid sample and begin again.

7. With a very slight up and down motion of the Temperature Probe, *continuously* stir the lauric acid during the cooling. Hold the top of the probe and *not* its wire.

8. Continue with the experiment until data collection has stopped after 10 minutes. Use the hot water bath provided by your teacher to melt the probe out of the solid lauric acid. *Do not* attempt to pull the probe out—this might damage it. Carefully wipe any excess lauric acid liquid from the probe with a paper towel or tissue. Return the test tube containing lauric acid to the place directed by your teacher.

9. To determine the freezing temperature of pure lauric acid, you need to determine the mean (or average) temperature in the portion of graph with nearly constant temperature. Move the mouse pointer to the beginning of the graph's flat part. Press the mouse button and hold it down as you drag across the flat part of the curve, selecting only the points in the plateau. Click the Statistics button, 〈ᵛ²ₛₜₐₜ〉. The mean temperature value for the selected data is listed in the statistics box on the graph. Record this value as the freezing temperature of lauric acid. Close the statistics box.

Part II Freezing Temperature of a Solution of Benzoic Acid and Lauric Acid

10. Store your data by choosing Store Latest Run from the Experiment menu. Hide the curve from your first run by clicking on the vertical axis label and unchecking the appropriate box. Click ⸂ OK ⸃.

11. Obtain a test tube containing a melted solution with ~1 g of benzoic acid dissolved in ~8 g of lauric acid. Record the *precise masses* of benzoic acid and lauric acid as indicated on the label of the test tube. Repeat Steps 3-8 to determine the freezing point of this mixture.

12. When you have completed Step 8, click on the Examine button, 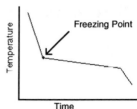. To determine the freezing point of the benzoic acid-lauric acid solution, you need to determine the temperature at which the mixture initially started to freeze. Unlike pure lauric acid, cooling a mixture of benzoic acid and lauric acid results in a gradual linear decrease in temperature during the time period when freezing takes place. As you move the mouse cursor across the graph, the temperature (y) and time (x) data points are displayed in the examine box on the graph. Locate the initial freezing temperature of the solution, as shown here. Record the freezing point in your data table.

13. To print a graph of temperature *vs.* time showing both data runs:

    a. Click on the vertical-axis label of the graph. To display both temperature runs, click More, and check the Run 1 and Latest Temperature boxes. Click [OK].

    b. Label both curves by choosing Text Annotation from the Insert menu, and typing "Lauric acid" (or "Benzoic acid-lauric acid mixture") in the edit box. Then drag each box to a position on or near its respective curve.

    c. Print the graph.

PROCESSING THE DATA

1. Determine the difference in freezing temperatures, Δt, between the pure lauric acid (t_1) and the mixture of lauric acid and benzoic acid (t_2). Use the formula, $\Delta t = t_1 - t_2$.

2. Calculate molality (m), in mol/kg, using the formula, $\Delta t = K_f \cdot m$ ($K_f = 3.9°C$-kg/mol for lauric acid).

3. Calculate moles of benzoic acid solute, using the answer in Step 2 (in mol/kg) and the mass (in kg) of lauric acid solvent.

4. Calculate the *experimental* molecular weight of benzoic acid, in g/mol. Use the original mass of benzoic acid from your data table, and the moles of benzoic acid you found in the previous step.

5. Determine the *accepted* molecular weight for benzoic acid from its formula, C_6H_5COOH.

6. Calculate the percent error.

EXTENSION

Here is another method that can be used to determine the freezing temperature from your data in Part II. With a graph of the Part II data displayed, use this procedure:

1. Move the mouse pointer to the initial part of the cooling curve, where the temperature has an initial rapid decrease (before freezing occurred). Press the mouse button and hold it down as you drag across the linear region of this steep temperature decrease.

2. Click the Linear Fit button, ⬚.

3. Now press the mouse button and drag over the next linear region of the curve (the gently sloping section of the curve where freezing took place). Press the mouse button and hold it down as you drag only this linear region of the curve.

4. Click ⊡ again. The graph should now have two regression lines displayed.

5. Choose Interpolate from the Analyze menu. Move the mouse pointer left to the point where the two regression lines intersect. When the small circles on each cursor line overlap each other at the intersection, the temperatures shown in either examine box should be equal to the freezing temperature for the benzoic acid-lauric acid mixture.

DATA AND CALCULATIONS

Mass of lauric acid	g
Mass of benzoic acid	g
Freezing temperature of pure lauric acid	°C
Freezing point of the benzoic acid–lauric acid mixture	°C

Freezing temperature depression, Δt	°C
Molality, m	mol/kg
Moles of benzoic acid	mol
Molecular weight of benzoic acid (experimental)	g/mol
Molecular weight of benzoic acid (accepted)	g/mol
Percent error	%

Using Freezing Point Depression to Find Molecular Weight

1. The student pages with complete instructions for data-collection using LabQuest App, Logger *Pro* (computers), EasyData or DataMate (calculators), and DataPro (Palm handhelds) can be found on the CD that accompanies this book. See *Appendix A* for more information.

2. Lauric acid, $CH_3(CH_2)_{10}COOH$, has an accepted melting point of 44.0°C. It is also called dodecanoic acid. It has an accepted molecular weight value of 200.32.

3. Do not use the white TI Temperature Probe that was shipped with the original CBL; using it in lauric acid or benzoic acid-lauric acid can permanently damage the probe.

4. The ΔT value limits the accuracy of the final answer for molecular weight to two significant figures. As a result, the final molecular weight in the sample data was expressed as 120 g/mol, not 119 g/mol.

5. Test tube sizes 19×150 mm, 20×150 mm, and 25×150 mm work well.

6. In order to have plenty of time to run two trials in one 40–50 minute period, we recommend that you have pre-measured lauric acid and lauric acid-benzoic acid samples in test tubes. In Part I, a sample of lauric acid that is approximately 8 grams (filled to a depth of about 10 cm) works well. In Part II, measure 8 g of lauric acid and 1 g of benzoic acid into each test tube, to the precision of your electronic balance (at least 0.01 g). Place a label on each test tube showing the precise masses of the lauric acid and the benzoic acid. Be sure the label will be above the water level of the water bath. Students will record and use these masses in their calculations.

7. A hot-water bath on a hot plate adjusted to a low setting (one that maintains the water bath at a temperature of 80–90°C) can be used to melt the samples and maintain them in the liquid state. Place the beaker with samples of lauric acid on one hot plate at one location, and the beaker with samples of the lauric acid-benzoic acid mixtures on a second hot plate at another location.

8. If you follow the procedure in Steps 4 and 5 above, you can easily reuse samples of lauric acid and the lauric acid-benzoic acid mixtures. There should be very little contamination of the samples if you remind students to wipe any excess liquid from the probe at the end of a trial. This method also makes clean-up a great deal easier for you at the end of the lab. Stopper the test tubes and store them for future use.

9. Step 6 mentions at least one temperature reading of 50°C or greater. This ensures a portion of the graph showing the drop down to the 44°C freezing temperature of lauric acid.

10. Have a hot-water bath available to free probes that have become frozen in lauric acid (or the mixture). The test tubes can be immersed in the bath to melt the solution and free the probes.

11. If you have class periods longer than 50 minutes, you may choose to have students weigh out their own samples of lauric acid and benzoic acid. We still recommend having hot plates with hot water baths available, to save students the time to heat up a water bath.

12. **HAZARD ALERT:**

Benzoic acid: Moderately toxic by ingestion; irritates eyes, skin and respiratory tract; combustible. Hazard Code: C—Somewhat hazardous.

The hazard information reference is: Flinn Scientific, Inc., *Chemical & Biological Catalog Reference Manual*, (800) 452-1261, www.flinnsci.com. See *Appendix D* of this book, *Chemistry with Vernier*, for more information.

11. The stored calibrations for the Stainless Steel Temperature Probe works well for this experiment.

SAMPLE RESULTS

Mass of lauric acid	8.01 g
Mass of benzoic acid	1.00 g
Freezing temperature of pure lauric acid	44.0°C
Freezing point of the benzoic acid–lauric acid mixture	39.9°C

Freezing temperature depression, Δt	$\Delta t = 44.0°C - 39.9°C =$ 4.1°C
Molality, m	$m = \Delta t/K_f = \dfrac{4.1°C}{3.9°C\text{-kg/mol}} =$ 1.05 mol/kg
Moles of benzoic acid	$(1.05 \text{ mol/kg})(0.00801 \text{ kg}) =$ 0.00841 mol
Molecular weight of benzoic acid (experimental)	$\dfrac{1.00 \text{ g}}{0.00841 \text{ mol}} = 119 =$ 120 g/mol
Molecular weight of benzoic acid (accepted)	C: 7 x 12.0 = 84.0 H: 6 x 1.0 = 6.0 O: 2 x 16.0 = 32.0 122.0 122 g/mol
Percent error	$\dfrac{|122-120|}{|122|} \times 100 =$ 2 %

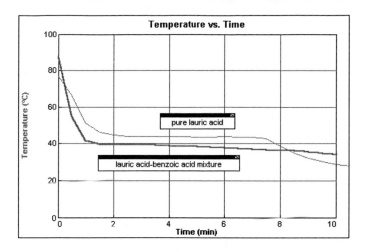

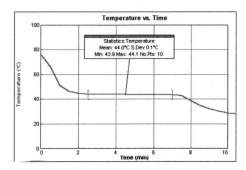

Pure lauric acid

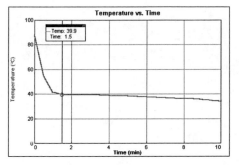

Lauric acid - benzoic acid mixture

	Time minutes	Temp °C		Time minutes	Temp °C
1	0.0	76.8	1	0.0	88.0
2	0.5	66.6	2	0.5	55.3
3	1.0	51.5	3	1.0	42.0
4	1.5	46.2	4	1.5	39.9
5	2.0	44.7	5	2.0	39.7
6	2.5	44.1	6	2.5	39.7
7	3.0	44.1	7	3.0	39.4
8	3.5	44.1	8	3.5	39.4
9	4.0	44.1	9	4.0	39.2
10	4.5	44.1	10	4.5	38.8
11	5.0	43.9	11	5.0	38.8
12	5.5	43.9	12	5.5	38.4
13	6.0	44.1	13	6.0	38.2
14	6.5	43.9	14	6.5	37.7
15	7.0	43.9	15	7.0	37.1
16	7.5	43.0	16	7.5	36.7
17	8.0	38.6	17	8.0	36.5
18	8.5	34.8	18	8.5	36.5
19	9.0	32.4	19	9.0	35.8
20	9.5	30.5	20	9.5	35.0
21	10.0	29.0	21	10.0	34.1

Energy Content of Foods

All human activity requires "burning" food for energy. In this experiment, you will determine the energy released (in kJ/g) as various foods, such as cashews, marshmallows, peanuts, and popcorn, burn. You will look for patterns in the amounts of energy released during burning of the different foods.

OBJECTIVES

In this experiment, you will

- Determine the energy released from various foods as they burn.
- Look for patterns in the amounts of energy released during burning of different foods.

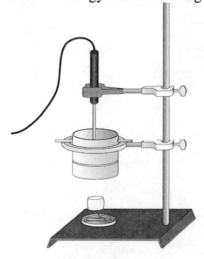

Figure 1

MATERIALS

computer	utility clamp
Vernier computer interface	2 stirring rods
Logger*Pro*	ring stand and 10 cm ring
Temperature Probe	100 mL graduated cylinder
two food samples	small can
food holder	cold water
wooden splint	matches

PROCEDURE

1. Obtain and wear goggles.

2. Obtain a piece of one of the two foods assigned to you and a food holder like the one shown in Figure 1. Find and record the initial mass of the food sample and food holder. **CAUTION:** *Do not eat or drink in the laboratory.*

3. Determine and record the mass of an empty can. Add 50 mL of cold water to the can. Obtain the cold water from your teacher. Determine and record the mass of the can and water.

4. Set up the apparatus as shown in Figure 1. Use a ring and stirring rod to suspend the can about 2.5 cm (1 inch) above the food sample. Use a utility clamp to suspend the Temperature Probe in the water. The probe should not touch the bottom of the can. Remember: The Temperature Probe must be in the water for at least 30 seconds before you do Step 6.

5. Connect the probe to the computer interface. Prepare the computer for data collection by opening the file "16 Energy of Foods" from *Chemistry with Vernier* folder of Logger *Pro*.

6. Click ▸Collect to begin measuring temperature. Record the initial temperature. Remove the food sample from under the can and use a wooden splint to light it. Quickly place the burning food sample directly under the center of the can. Allow the water to be heated until the food sample stops burning. **CAUTION:** *Keep hair and clothing away from open flames.*

7. Continue stirring the water until the temperature stops rising. Record this final temperature. Click ■ Stop to end data collection.

8. Determine and record the final mass of the food sample and food holder.

9. Examine the initial readings in the table to confirm the initial temperature, t_1. To confirm the final temperature, t_2, click the Statistics button, STAT. The maximum temperature is listed in the statistics box on the graph.

10. Repeat the procedure for the second food sample. Use a new 50 mL portion of cold water.

11. When you are finished, place burned food, used matches, and partially-burned wooden splints in the container provided by the teacher.

PROCESSING THE DATA

1. Find the mass of water heated for each sample.

2. Find the change in temperature of the water, Δt, for each sample.

3. Calculate the heat absorbed by the water, q, using the equation

$$q = C_p \cdot m \cdot \Delta t$$

where q is heat, C_p is the specific heat capacity, m is the mass of water, and Δt is the change in temperature. For water, C_p is 4.18 J/g°C. Change your final answer to kJ.

4. Find the mass (in g) of each food sample burned.

5. Use the results of Step 3 and 4 to calculate the energy content (in kJ/g) of each food sample.

6. Record your results and the results of other groups in the Class Results Table. Which food had the highest energy content? The lowest energy content?

7. Food energy is often expressed in a unit called a Calorie. There are 4.18 kJ in one Calorie. Based on the class average for peanuts, calculate the number of Calories in a 50 g package of peanuts.

8. Two of the foods in the experiment have a high fat content (peanuts and cashews) and two have a high carbohydrate content (marshmallows and popcorn). From your results, what generalization can you make about the relative energy content of fats and carbohydrates?

DATA AND CALCULATIONS

Food type		
Initial mass of food and holder	g	g
Final mass of food and holder	g	g
Mass of food burned	g	g
Mass of can and water	g	g
Mass of empty can	g	g
Final temperature, t_2	°C	°C
Initial temperature, t_1	°C	°C
Temperature change, Δt	°C	°C

Heat, q	kJ	kJ
Energy content in kJ/g	kJ/g	kJ/g

CLASS RESULTS TABLE

Marshmallows	Peanuts	Cashews	Popcorn
kJ/g	kJ/g	kJ/g	kJ/g
kJ/g	kJ/g	kJ/g	kJ/g
kJ/g	kJ/g	kJ/g	kJ/g
kJ/g	kJ/g	kJ/g	kJ/g
kJ/g	kJ/g	kJ/g	kJ/g

Average for each food type

kJ/g	kJ/g	kJ/g	kJ/g

TEACHER INFORMATION

Energy Content of Foods

1. The student pages with complete instructions for data-collection using LabQuest App, Logger *Pro* (computers), EasyData or DataMate (calculators), and DataPro (Palm handhelds) can be found on the CD that accompanies this book. See *Appendix A* for more information.

2. The food stand can be made using an extra-large paper clip and a small jar lid, such as a baby-food jar lid. Partially straighten the paper clip, then bend a small loop at one end. This loop will cradle food samples. Bend the other end to a "V" shape—this will be the base. Glue the paper clip into the lid. An advantage of such a stand is its ability to catch pieces of burned food that fall.

3. Small soup cans work well. Remove the paper and the top. Place two holes, large enough to accommodate a stirring rod, near the top. Be sure to remove all sharp edges. Many teachers prefer to use aluminum beverage cans.

4. Beginning with water cooled to 15–18°C gives good results.

5. We use cashews, marshmallows, peanuts, and popcorn (unbuttered) for this experiment. There are other good choices. Use of nuts, especially peanuts, is being restricted and phased out of schools due to increasing numbers of allergic reactions and the heightened sensitivity some students exhibit.

6. Because peanuts and cashew nuts release very large amounts of heat as they burn, you may want to have your students use 100 mL portions of cold water when doing these foods.

7. The stored calibrations for the Stainless Steel Temperature Probe works well for this experiment.

ANSWERS TO QUESTIONS

6. Cashews and peanuts have the highest energy content. Marshmallows and popcorn have the lowest.

7. Calories in a 50.0 g package of peanuts: (12.0 kJ/g)(50.0 g)(1 Cal / 4.18 kJ) = 144 Cal

8. The two foods with a high fat content, cashews and peanuts, had a much higher energy content than those with a high carbohydrate content (nearly double the energy content).

SAMPLE RESULTS

Food type	Cashews
Initial mass of food and holder	14.04 g
Final mass of food and holder	13.36 g
Mass of food burned	0.68 g
Mass of can and water	90.69 g
Mass of empty can	41.31 g
Mass of water heated	49.38 g
Final temperature, t_2	52.9°C
Initial temperature, t_1	15.4°C
Temperature change, Δt	37.5°C

Heat, q	$(4.18 \text{ J/g°C})(49.38 \text{ g})(37.5°C)$ $7740 \text{ J} = 7.74 \text{ kJ}$
Energy content in kJ/g	$\dfrac{7.74 \text{ kJ}}{0.68 \text{ g}} =$ 11.4 kJ/g

CLASS AVERAGES

Marshmallows	Peanuts	Cashews	Popcorn
4.2–5.8 kJ/g	11.0–12.5 kJ/g	11.0–12.0 kJ/g	5.0–8.4 kJ/g

Energy Content of Fuels

In this experiment, you will find and compare the heat of combustion of two different fuels: paraffin wax and ethanol. Paraffin is a member of a group of compounds called alkanes that are composed entirely of carbon and hydrogen atoms. Many alkanes, such as gasoline and diesel oil, are important fuels. Ethanol, C_2H_5OH, is used as a gasoline additive (gasohol) and as a gasoline substitute. In this experiment, you will compare the energy content of paraffin and ethanol by measuring their heats of combustion in kJ/g of fuel.

In order to find the heat of combustion, you will first use the energy from burning ethanol or paraffin to heat a known quantity of water. By monitoring the temperature of the water, you can find the amount of heat transferred to it, using the formula

$$q = C_p \cdot m \cdot \Delta t$$

where q is heat, C_p is the specific heat capacity of water, m is the mass of water, and Δt is the change in temperature of the water. Finally, the *amount* of fuel burned will be taken into account by calculating the heat *per gram* of fuel consumed in the combustion.

OBJECTIVES

In this experiment, you will

- Compare the heat of combustion for paraffin wax and ethanol.
- Calculate the heat of combustion and percent efficiency for both fuels.

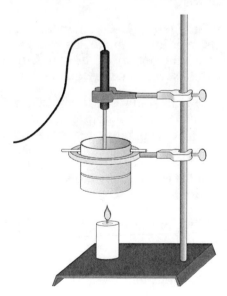

Figure 1

MATERIALS

computer
Vernier computer interface
Logger*Pro*
Temperature Probe
candle
alcohol burner with ethanol
7.5 × 12.5 cm (3" × 5") index card

utility clamp
2 stirring rods
ring stand and 10 cm (4 inch) ring
100 mL graduated cylinder
small can
cold water
matches

PROCEDURE

1. Obtain and wear goggles.

2. Connect the probe to the computer interface. Prepare the computer for data collection by opening the file "17 Energy of Fuels" from the *Chemistry with Vernier* folder.

Part 1 Paraffin Wax

3. Prepare a candle for the experiment by holding a lighted match near its base, so that some melted wax falls onto a 3" × 5" (7.5 × 12.5 cm) index card. Immediately push the candle into the melted wax and hold it there for a moment to fasten it to the card.

4. Find and record the combined mass of the candle and index card (or the alcohol burner and contents in Part 2 below).

5. Determine and record the mass of an empty can. Add 100 mL of chilled water to the can (use 200 mL of chilled water in Part 2). Obtain the chilled water from your teacher. Determine and record the mass of the can and water.

6. Set up the apparatus as shown in Figure 1. Use a ring and stirring rod to suspend the can about 5 cm above the wick. Use a utility clamp to suspend the Temperature Probe in the water. The probe should not touch the bottom of the can. **Important:** The Temperature Probe must be in the water for at least 30 seconds before you do Step 7.

7. Click ▶Collect to begin measuring temperature. After the computer has taken readings for about 30 seconds, light the candle (or alcohol burner in Part 2). Record the initial temperature. Heat the water until its temperature reaches 40°C and then extinguish the flame. **CAUTION:** *Keep hair and clothing away from an open flame.*

8. Continue stirring the water until the temperature stops rising. Record this final temperature. Click ■Stop to end data collection.

9. Determine and record the final mass of the cooled candle and index card, including all drippings (or the cooled alcohol burner and contents in Part 2).

10. Examine the initial readings in the table to confirm the initial temperature, t_1. To confirm the final temperature, t_2, click the Statistics button, ⌐√⌐. The maximum temperature is listed in the statistics box on the graph.

Part 2 Ethanol

11. Repeat Steps 4–10 using ethanol in an alcohol burner. Be sure to use 200 mL of chilled water in Step 5.

PROCESSING THE DATA

1. Find the mass of water heated.

2. Find the change in temperature of the water, Δt.

3. Calculate the heat absorbed by the water, q, using the formula in the introduction of this experiment. For water, C_p is 4.18 J/g°C. Change your final answer to kJ.

4. Find the mass of paraffin or ethanol burned.

5. Calculate the heat of combustion for paraffin and ethanol, in kJ/g. Use your Step 3 and Step 4 answers.

6. Calculate the *% efficiency* in both trials of the experiment. Divide your experimental value (in kJ/g) by the accepted value, and multiply the answer by 100. The accepted heat of combustion of paraffin is 41.5 kJ/g, and for ethanol the value is 30.0 kJ/g.

7. Based on your results, which fuel produces more energy per gram burned? Give an explanation for the difference. (Hint: Ethanol, C_2H_5OH, is an *oxygenated* molecule; paraffin, $C_{25}H_{52}$, does not contain oxygen.)

8. Suggest some advantages of using ethanol (or paraffin) as a fuel.

9. Discuss heat loss factors that contribute to the inefficiency of the experiment.

DATA AND CALCULATIONS

	Part 1: Paraffin	Part 2: Ethanol
Initial mass of fuel + container	g	g
Final mass of fuel + container	g	g
Mass of fuel burned	g	g
Mass of can and water	g	g
Mass of empty can	g	g
Final temperature, t_2	°C	°C
Initial temperature, t_1	°C	°C
Temperature change, Δt	°C	°C
Heat, q	kJ	kJ
Heat of combustion, in kJ/g	kJ/g paraffin	kJ/g ethanol
% efficiency	%	%

Energy Content of Fuels

1. The student pages with complete instructions for data-collection using LabQuest App, Logger *Pro* (computers), EasyData or DataMate (calculators), and DataPro (Palm handhelds) can be found on the CD that accompanies this book. See *Appendix A* for more information.

2. Alcohol burners, available from science supply companies, work well with ethanol. Because alcohol evaporates rapidly, the mass of the alcohol and burner should be determined immediately before and after use.

3. Small soup cans work well. Completely remove the top and the label. Place two holes, large enough to fit a stirring rod, near the top. Be sure to remove all sharp edges. Many teachers prefer to use aluminum beverage cans.

4. Have ice water available and make sure all ice is removed from the water to be used. Water initially at 4–5°C gives best results, because starting 17–19°C below and finishing 17–19°C above room temperature tends to equalize heat exchange with the room.

5. Having the candles mounted on a 3" × 5" (7.5 × 12.5 cm) index card will catch candle drippings.

6. You may wish to have your students calculate the *molar* heat of combustion for these fuels. If so, use the formula $C_{25}H_{52}$ for paraffin (352 g/mol) and C_2H_5OH for ethanol (46 g/mol). Using the experimental values in the sample data, the values obtained are:

    paraffin: (24.4 kJ/g)(352 g/mol) = 8590 kJ/mol   (accepted value = 14600 kJ/mol)

    ethanol: (16.7 kJ/g)(46 g/mol) = 768 kJ/mol   (accepted value = 1380 kJ/mol)

    We use kJ/g in this experiment, because fuels are usually compared using a per-gram value for heat of combustion. However, it is also interesting to compare these two fuels using the same number of *moles* (and the same number of molecules).

7. The stored calibrations for the Stainless Steel Temperature Probe works well for this experiment.

8. Liquid Sterno® (also called Canned Heat®) can be used in place of alcohol burners. The smaller size, 2–5/8 oz., works well for this experiment. Each can burns about two hours. Have students place the water can about 2.5 cm above the Sterno can. The directions on the Sterno can recommend igniting it by touching a match to the contents. To extinguish, slide the cover upside down on top of the can (wait for the can to cool before replacing the cover).

    The gel-like structure of Sterno that traps ethanol fuel is formed by the saponification reaction between stearic acid and sodium hydroxide. According to the label, Sterno also contains 3.3% methanol. Starting with initial water temperatures of 6–7°C, 200 g of water will be heated to a temperature of 41–44°C in 5 minutes. We get heat of combustion values in the range of 8.5–10.5 kJ/g.

ANSWERS TO QUESTIONS

7. Paraffin's experimental heat of combustion (23.1 kJ/g) was greater than the value for ethanol (16.7 kJ/g). Ethanol can be considered to be a partially combusted molecule; oxygen is already included in its structure. Therefore, less energy should be produced when it is burned.

8. Ethanol is a renewable resource; paraffin is a petroleum product and is not renewable. Ethanol is a liquid at room temperature; liquids are needed for use in automobile engines and many furnaces. Paraffin produces more energy per gram of fuel.

9. A major factor contributing to the inefficiency of this experimental procedure is loss of heat to the air.

SAMPLE RESULTS

	Part 1: Paraffin	Part 2: Ethanol
Initial mass of fuel+container	24.82 g	84.07 g
Final mass of fuel+container	24.18 g	82.69 g
Mass of fuel burned	0.64 g	1.38 g
Mass of can and water	139.96 g	236.42 g
Mass of empty can	40.44 g	36.55 g
Mass of water heated	99.52 g	199.87 g
Final temperature, t_2	41.2°C	33.9°C
Initial temperature, t_1	5.6°C	6.5°C
Temperature change, Δt	35.6°C	27.4°C

Heat, q	(4.18 J/g°C)(99.52 g)(35.6°C) 14810 J = 14.8 kJ	(4.18 J/g°C)(199.87 g)(35.6°C) 14810 J = 14.8 kJ
Heat of combustion, in kJ/g	$\dfrac{14.8 \text{ kJ}}{0.64 \text{ g}}$ = 23.1 kJ/g paraffin	$\dfrac{23.0 \text{ kJ}}{1.38 \text{ g}}$ = 16.7 kJ/g ethanol
% efficiency	$\dfrac{23.1 \text{ kJ/g}}{41.5 \text{ kJ/g}}$ × 100 = 55.7 %	$\dfrac{16.7 \text{ kJ/g}}{30.0 \text{ kJ/g}}$ × 100 = 55.7 %

Additivity of Heats of Reaction: Hess's Law

In this experiment, you will use a Styrofoam-cup calorimeter to measure the heat released by three reactions. One of the reactions is the same as the combination of the other two reactions. Therefore, according to Hess's law, the heat of reaction of the one reaction should be equal to the sum of the heats of reaction for the other two. This concept is sometimes referred to as the *additivity of heats of reaction*. The primary objective of this experiment is to confirm this law. The reactions we will use in this experiment are:

(1) Solid sodium hydroxide dissolves in water to form an aqueous solution of ions.

$$NaOH(s) \longrightarrow Na^+(aq) + OH^-(aq) \quad \Delta H_1 = ?$$

(2) Solid sodium hydroxide reacts with aqueous hydrochloric acid to form water and an aqueous solution of sodium chloride.

$$NaOH(s) + H^+(aq)) + Cl^-(aq) \longrightarrow H_2O(l) + Na^+(aq) + Cl^-(aq) \quad \Delta H_2 = ?$$

(3) Solutions of aqueous sodium hydroxide and hydrochloric acid react to form water and aqueous sodium chloride.

$$Na^+(aq) + OH^-(aq) + H^+(aq)) + Cl^-(aq) \longrightarrow H_2O(l) + Na^+(aq) + Cl^-(aq) \quad \Delta H_3 = ?$$

OBJECTIVES

In this experiment, you will

- Combine equations for two reactions to obtain the equation for a third reaction.
- Use a calorimeter to measure the temperature change in each of three reactions.
- Calculate the heat of reaction, ΔH, for the three reactions.
- Use the results to confirm Hess's law.

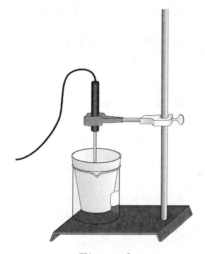

Figure 1

You will use a Styrofoam cup in a beaker as a calorimeter, as shown in Figure 1. For purposes of this experiment, you may assume that the heat loss to the calorimeter and the surrounding air is negligible. Even if heat is lost to either of these, it is a fairly constant factor in each part of the experiment, and has little effect on the final results.

PRE-LAB EXERCISE

In the space below, combine two of the above equations algebraically to obtain the third equation. Indicate the number of each reaction on the shorter lines.

_____   _____

_____   _____

_____   _____

MATERIALS

computer	100 mL of water
Vernier computer interface	4.00 g of solid NaOH
Logger *Pro*	ring stand
Temperature Probe	utility clamp
50 mL of 1.0 M NaOH	stirring rod
50 mL of 1.0 M HCl	Styrofoam cup
100 mL of 0.50 M HCl	250 mL beaker

PROCEDURE

Reaction 1

1. Obtain and wear goggles.

2. Connect the probe to the computer interface. Prepare the computer for data collection by opening the file "18 Hess's Law" from the *Chemistry with Vernier* folder.

3. Place a Styrofoam cup into a 250 mL beaker as shown in Figure 1. Measure out 100.0 mL of water into the Styrofoam cup. Lower the Temperature Probe into the solution.

4. Use a utility clamp to suspend a Temperature Probe from a ring stand as shown in Figure 1.

5. Weigh out about 2 grams of solid sodium hydroxide, NaOH, and record the mass to the nearest 0.01 g. Since sodium hydroxide readily picks up moisture from the air, it is necessary to weigh it and proceed to the next step without delay. **CAUTION:** *Handle the NaOH and resulting solution with care.*

6. Click on ▶ Collect to begin data collection and obtain the initial temperature, t_1. It may take several seconds for the Temperature Probe to equilibrate at the temperature of the solution. After three or four readings at the same temperature have been obtained, add the solid NaOH to the Styrofoam cup. Using the stirring rod, stir continuously for the remainder of the 200 seconds or until the temperature maximizes. As soon as the temperature has begun to drop after reaching a maximum, you may terminate the trial by clicking ■ Stop .

7. Examine the initial readings in the table to determine the initial temperature, t_1. To determine the final temperature, t_2, click the Statistics button, ⬚. The maximum temperature is listed in the statistics box on the graph. Record t_1 and t_2 in your data table.

8. Rinse and dry the Temperature Probe, Styrofoam cup, and stirring rod. Dispose of the solution as directed by your instructor.

Reaction 2

9. Repeat Steps 3–8 using 100.0 mL of 0.50 M hydrochloric acid, HCl, instead of water. **CAUTION:** *Handle the HCl solution and NaOH solid with care.*

Reaction 3

10. Repeat Steps 3–8, initially measuring out 50.0 mL of 1.0 M HCl (instead of water) into the Styrofoam calorimeter. In Step 5, instead of solid NaOH, measure 50.0 mL of 1.0 M NaOH solution into a graduated cylinder. After t_1 has been determined for the 1.0 M HCl, add the 1.0 M NaOH solution to the Styrofoam cup. **CAUTION:** *Handle the HCl and NaOH solutions with care.*

PROCESSING DATA

1. Determine the mass of 100 mL of solution for each reaction (assume the density of each solution is 1.00 g/mL).

2. Determine the temperature change, Δt, for each reaction.

3. Calculate the heat released by each reaction, q, by using the formula:

$$q = C_p \cdot m \cdot \Delta t \quad (C_p = 4.18 \text{ J/g}°\text{C})$$

Convert joules to kJ in your final answer.

4. Find ΔH ($\Delta H = -q$).

5. Calculate moles of NaOH used in each reaction. In Reactions 1 and 2, this can be found from the mass of the NaOH. In Reaction 3, it can be found using the molarity, M, of the NaOH and its volume, in L.

6. Use the results of the Step 4 and Step 5 calculations to determine ΔH/mol NaOH in each of the three reactions.

7. To verify the results of the experiment, combine the heat of reaction (ΔH/mol) for Reaction 1 and Reaction 3. This sum should be similar to the heat of reaction (ΔH/mol) for Reaction 2. Using the value in Reaction 2 as the accepted value and the sum of Reactions 1 and 3 as the experimental value, find the percent error for the experiment.

DATA AND CALCULATIONS

	Reaction 1	Reaction 2	Reaction 3
1. Mass of solid NaOH	g	g	(no solid NaOH mass)
2. Mass (total) of solution	g	g	g
3. Final temperature, t_2	°C	°C	°C
4. Initial temperature, t_1	°C	°C	°C
5. Change in temperature, Δt	°C	°C	°C
6. Heat, q	kJ	kJ	kJ
7. ΔH	kJ	kJ	kJ
8. Moles of NaOH	mol	mol	mol
9. ΔH/mol	kJ/mol	kJ/mol	kJ/mol
10. Experimental value			kJ/mol
11. Accepted value			kJ/mol
12. Percent error			%

TEACHER INFORMATION

Additivity of Heats of Reaction: Hess's Law

1. The student pages with complete instructions for data-collection using LabQuest App, Logger *Pro* (computers), EasyData or DataMate (calculators), and DataPro (Palm handhelds) can be found on the CD that accompanies this book. See *Appendix A* for more information.

2. Preparation of solutions:

    0.5 M HCl (42.8 mL of concentrated HCl per 1 L solution) **HAZARD ALERT:** Highly toxic by ingestion or inhalation; severely corrosive to skin and eyes. Hazard Code: A—Extremely hazardous.

    1.0 M HCl (85.6 mL of concentrated HCl per 1 L solution) **HAZARD ALERT:** Highly toxic by ingestion or inhalation; severely corrosive to skin and eyes. Hazard Code: A—Extremely hazardous.

    1.0 M NaOH (40.0 g of solid NaOH per 1 L solution) **HAZARD ALERT:** Corrosive solid; skin burns are possible; much heat evolves when added to water; very dangerous to eyes; wear face and eye protection when using this substance. Wear gloves. Hazard Code: B—Hazardous.

    The hazard information reference is: Flinn Scientific, Inc., *Chemical & Biological Catalog Reference Manual*, (800) 452-1261, www.flinnsci.com. See *Appendix D* of this book, *Chemistry with Vernier*, for more information.

2. It is very important to prepare the solutions at least one day in advance so they will be at room temperature prior to doing the experiment.

3. You should discuss with your students three assumptions made in this lab. One is that the specific heat capacity, C_p, for the aqueous solutions is assumed to be the same that of pure water, 4.18 J/g°C. They are, in fact, very nearly the same. The second assumption is that the density of the aqueous solutions is 1.00 g/mL. Since this is very nearly the case, we can use a mass of 100 g for 100 mL of solution. The procedure for Reaction 3 uses the assumption that initial HCl solution and NaOH solution temperatures are the same. If you make up the solutions at least one day in advance and store them together, the two temperatures will be the same or nearly the same.

4. The stored calibrations for the Stainless Steel Temperature Probe works well for this experiment.

PRE-LAB EXERCISE ANSWER

Reaction 2 is a combination of Reactions 1 and 3

(1) $NaOH(s) \longrightarrow Na^+(aq) + OH^-(aq)$

(3) $Na^+(aq) + OH^-(aq) + H^+(aq) + Cl^-(aq) \longrightarrow H_2O(1) + Na^+(aq) + Cl^-(aq)$

(2) $NaOH(s) + H^+(aq) + Cl^-(aq) \longrightarrow H_2O(1) + Na^+(aq) + Cl^-(aq)$

SAMPLE RESULTS

	Reaction 1	Reaction 2	Reaction 3
1. Mass of solid NaOH	2.00 g	1.92 g	(no solid NaOH mass)
2. Mass (total) of solution	100.0 g	100.0 g	100.0 g
3. Final temperature, t_2	27.0°C	33.8°C	29.1°C
4. Initial temperature, t_1	22.0°C	22.2°C	22.3°C
5. Change in temperature, Δt	5.0°C	11.6°C	6.8°C
6. Heat, q	(4.18)(100 g)(5.0°C) = 2090 J = 2.09 kJ	(4.18)(100 g)(11.6°C) = 4850 J = 4.85 kJ	(4.18)(100 g)(6.8°C) = 2840 J = 2.84 kJ
7. ΔH	$\Delta H = -q = -(2.09 \text{ kJ})$ −2.09 kJ	$\Delta H = -q = -(4.85 \text{ kJ})$ −4.85 kJ	$\Delta H = -q = -(2.84 \text{ kJ})$ −2.84 kJ
8. Moles of NaOH	(2.00 g)(1 mol/40 g) = 0.0500 mol	(1.92 g)(1 mol/40 g) = 0.0480 mol	(1 mol/L)(0.050 L) = 0.0500 mol
9. ΔH/mol	$\dfrac{-2.09 \text{ kJ}}{0.0500 \text{ mol}} =$ −41.8 kJ/mol	$\dfrac{-4.85 \text{ kJ}}{0.0480 \text{ mol}} =$ −101 kJ/mol	$\dfrac{-2.84 \text{ kJ}}{0.0500 \text{ mol}} =$ −56.8 kJ/mol
10. Experimental value kJ/mol (Reaction 1 + Reaction 3) −41.8 kJ + (−56.8 kJ) = −98.6 kJ/mol			
11. Accepted value kJ/mol (Reaction 2)			−101 kJ/mol
12. Percent error	$\dfrac{\lvert -101-(-98.6)\rvert}{\lvert -101 \rvert}$ **x** 100 =		2.4 %

Heat of Combustion: Magnesium

In Experiment 18, you learned about the additivity of reaction heats as you confirmed Hess's Law. In this experiment, you will use this principle as you determine a heat of reaction that would be difficult to obtain by direct measurement—the heat of combustion of magnesium ribbon. The reaction is represented by the equation

(4) $Mg(s) + 1/2\ O_2(g) \longrightarrow MgO(s)$

This equation can be obtained by combining equations (1), (2), and (3):

(1) $MgO(s) + 2\ HCl(aq) \longrightarrow MgCl_2(aq) + H_2O(l)$

(2) $Mg\ (s) + 2\ HCl(aq) \longrightarrow MgCl_2(aq) + H_2\ (g)$

(3) $H_2(g) + 1/2\ O_2(g) \longrightarrow H_2O(l)$

The pre-lab portion of this experiment requires you to combine equations (1), (2), and (3) to obtain equation (4) before you do the experiment. Heats of reaction for equations (1) and (2) will be determined in this experiment. As you may already know, ΔH for reaction (3) is -285.8 kJ.

OBJECTIVES

In this experiment, you will

- Combine three chemical equations to obtain a fourth.
- Use prior knowledge about the additivity of reaction heats.
- Determine the heat of combustion of magnesium ribbon.

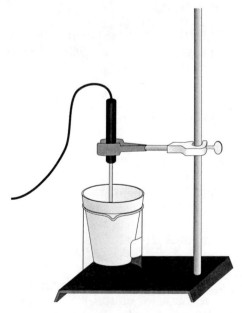

Figure 1

MATERIALS

computer	Styrofoam cup
Vernier computer interface	1.00 M HCl
Logger*Pro*	magnesium oxide, MgO
Temperature Probe	magnesium ribbon, Mg
ring stand	stirring rod
utility clamp	balance
100 mL graduated cylinder	weighing paper
250 mL beaker	

PRE-LAB EXERCISE

In the space provided below, combine equations (1), (2), and (3) to obtain equation (4).

(1) _____

(2) _____

(3) _____

(4) _____

PROCEDURE

1. Obtain and wear safety glasses and an apron.

2. Connect the probe to the computer interface. Prepare the computer for data collection by opening the file "19 Heat of Combustion" from the *Chemistry with Vernier* folder.

Reaction 1

3. Place a Styrofoam cup into a 250 mL beaker as shown in Figure 1. Measure out 100.0 mL of 1.00 M HCl into the Styrofoam cup. **CAUTION:** *Handle the HCl solution with care. It can cause painful burns if it comes in contact with the skin.*

4. Use a utility clamp and a slit stopper to suspend a Temperature Probe from a ring stand as shown in Figure 1. Lower the Temperature Probe into the solution in the Styrofoam cup.

5. Weigh out about 1.00 g of magnesium oxide, MgO, on a piece of weighing paper. Record the exact mass used in your data table. **CAUTION:** *Avoid inhaling magnesium oxide dust.*

6. Click ▶ **Collect** to begin data collection and obtain the initial temperature, t_1. After three or four readings at the same temperature (t_1) have been obtained, add the white magnesium oxide powder to the solution. Use a stirring rod to stir the cup contents until a maximum temperature has been reached and the temperature starts to drop. Click ■ **Stop** to end data collection.

7. Examine the initial readings in the table to determine the initial temperature, t_1. To determine the final temperature, t_2, click the Statistics button, $\boxed{\text{STAT}}$. The maximum temperature is listed in the statistics box on the graph. Record t_1 and t_2 in your data table.

8. Discard the solution as directed by your teacher.

Reaction 2

9. Repeat Steps 3–8 using about 0.50 g of magnesium ribbon rather than magnesium oxide powder. The magnesium ribbon has been pre-cut to the proper length by your teacher. Be sure to record the measured mass of the magnesium. **CAUTION:** *Do not breathe the vapors produced in the reaction!*

PROCESSING THE DATA

1. In the spaces provided, calculate the change in temperature, Δt, for Reactions 1 and 2.

2. Calculate the heat released by each reaction, q, using the formula

$$q = C_p \cdot m \cdot \Delta t$$

C_p = 4.18 J/g°C, and m = 100.0 g of HCl solution. Convert joules to kJ in your final answer.

3. Determine ΔH. ($\Delta H = -q$)

4. Determine the moles of MgO and Mg used.

5. Use your Step 3 and Step 4 results to calculate ΔH/mol for MgO and Mg.

6. Determine ΔH/mol Mg for Reaction 4. (Use your Step 5 results, your pre-lab work, and ΔH = –285.8 kJ for Reaction 3).

7. Determine the percent error for the answer you obtained in Step 6. The accepted value for this reaction can be found in a table of standard heats of formation.

DATA AND CALCULATIONS

	Reaction 1 (MgO)	Reaction 2 (Mg)
1. Volume of 1.00 M HCl	g	g
2. Final temperature, t_2	°C	°C
3. Initial temperature, t_1	°C	°C
4. Change in temperature, Δt	°C	°C
5. Mass of solid	g	g

6. Heat, q	kJ	kJ
7. ΔH	kJ	kJ
8. Moles	mol MgO	mol Mg
9. ΔH/mol	kJ/mol	kJ/mol

10. Determine ΔH/mol Mg for reaction (4)*.

    (1) _____ _____

    (2) _____ _____

    (3) _____ _____

    (4)* _____ _____

11. Percent error	kJ/mol

Heat of Combustion: Magnesium

1. The student pages with complete instructions for data-collection using LabQuest App, Logger *Pro* (computers), EasyData or DataMate (calculators), and DataPro (Palm handhelds) can be found on the CD that accompanies this book. See *Appendix A* for more information.

2. Have pre-cut strips of magnesium *ribbon* weighing about 0.50 g available. **HAZARD ALERT:** Flammable solid; burns with an intense flame; keep either dry sand or Flinn *Met-L-X*® available to use as a fire extinguisher. Hazard Code: C—Somewhat hazardous.

3. Each student or pair will need 200 mL of 1.00 M HCl (85.6 mL concentrated HCl reagent/liter). **HAZARD ALERT:** Highly toxic by ingestion or inhalation; severely corrosive to skin and eyes. Hazard Code: A—Extremely hazardous.

   Additional **HAZARD ALERT:** Magnesium oxide: Dust mildly toxic by inhalation. Hazard Code: C—Somewhat hazardous.

   The hazard information reference is: Flinn Scientific, Inc., *Chemical & Biological Catalog Reference Manual*, (800) 452-1261, www.flinnsci.com. See *Appendix D* of this book, *Chemistry with Vernier*, for more information.

4. The accepted value for ΔH is –602 kJ/mol Mg. You can supply this information to your students so they can calculate the percent error. This value may also be listed in a table of standard heats (enthalpies) of formation in your textbook.

5. The stored calibrations for the Stainless Steel Temperature Probe works well for this experiment.

SAMPLE RESULTS

	Reaction 1 (MgO)	Reaction 2 (Mg)
1. Volume of 1.00 M HCl	100.0 mL	100.0 mL
2. Final temperature, t_2	29.7°C	44.2°C
3. Initial temperature, t_1	22.3°C	22.4°C
4. Change in temperature, Δt	7.4°C	21.8°C
5. Mass of solid	1.00 g	0.50 g
6. Heat, q	(4.18 J/g°C)(100g)(7.4°C) = 3090 J = 3.09 kJ	(4.18 J/g°C)(100g)(21.8°C) = 9110 J = 9.11 kJ
7. ΔH	$\Delta H = -q = -(3.09 \text{ kJ}) =$ −3.09 kJ	$\Delta H = -q = -(9.11 \text{ kJ}) =$ −9.11 kJ
8. Moles	(1.00 g)(1 mol/40.3 g) = 0.0248 mol MgO	(0.50 g)(1 mol/24.3 g) = 0.0206 mol Mg
9. ΔH/mol	−3.09 kJ/0.0248 mol = −125 kJ/mol	−9.11 kJ/0.0206 mol = −442 kJ/mol

10. Determine ΔH/mol Mg for reaction (4)*

(1) $MgCl_2(aq) + H_2O(1) \longrightarrow MgO(s) + 2 HCl(aq)$ $\Delta H = +125$ kJ

(2) $Mg(s) + 2 HCl(aq) \longrightarrow MgCl_2(aq) + H_2(g)$ $\Delta H = -442$ kJ

(3) $H_2(g) + 1/2 O_2(g) \longrightarrow H_2O(1)$ $\Delta H = -286$ kJ

(4)*$Mg(s) + 1/2 O_2(g) \longrightarrow MgO(s)$ $\Delta H = -603$ kJ

11. Percent error	$\dfrac{\lvert -602-(-603) \rvert}{\lvert -602 \rvert} \times 100 =$	0.17 %

Chemical Equilibrium: Finding a Constant, K$_c$

The purpose of this lab is to experimentally determine the equilibrium constant, K$_c$, for the following chemical reaction:

$$Fe^{3+}(aq) + SCN^-(aq) \longleftrightarrow FeSCN^{2+}(aq)$$

iron(III)   thiocyanate   thiocyanoiron(III)

When Fe^{3+} and SCN^- are combined, equilibrium is established between these two ions and the $FeSCN^{2+}$ ion. In order to calculate K$_c$ for the reaction, it is necessary to know the concentrations of all ions at equilibrium: $[FeSCN^{2+}]_{eq}$, $[SCN^-]_{eq}$, and $[Fe^{3+}]_{eq}$. You will prepare four equilibrium systems containing different concentrations of these three ions. The equilibrium concentrations of the three ions will then be experimentally determined. These values will be substituted into the equilibrium constant expression to see if K_c is indeed constant.

You will use a Colorimeter or a Spectrometer to determine $[FeSCN^{2+}]_{eq}$. The $FeSCN^{2+}$ ion produces solutions with a red color. Because the red solutions absorb blue light very well, so Colorimeter users will be instructed to use the 470 nm (blue) LED. Spectrometer users will determine an appropriate wavelength based on the absorbance spectrum of the solution. The light striking the detector is reported as *absorbance* or *percent transmittance.* By comparing the absorbance of each equilibrium system, A_{eq}, to the absorbance of a *standard* solution, A_{std}, you can determine $[FeSCN^{2+}]_{eq}$. The standard solution has a known $FeSCN^{2+}$ concentration.

To prepare the standard solution, a very large concentration of Fe^{3+} will be added to a small initial concentration of SCN^- (hereafter referred to as $[SCN^-]_i$. The $[Fe^{3+}]$ in the standard solution is 100 times larger than $[Fe^{3+}]$ in the equilibrium mixtures. According to LeChatelier's principle, this high concentration forces the reaction far to the right, using up nearly 100% of the SCN^- ions. According to the balanced equation, for every one mole of SCN^- reacted, one mole of $FeSCN^{2+}$ is produced. Thus $[FeSCN^{2+}]_{std}$ is assumed to be equal to $[SCN^-]_i$.

Assuming $[FeSCN^{2+}]$ and absorbance are related directly (Beer's law), the concentration of $FeSCN^{2+}$ for any of the equilibrium systems can be found by:

$$[FeSCN^{2+}]_{eq} = \frac{A_{eq}}{A_{std}} \times [FeSCN^{2+}]_{std}$$

Knowing the $[FeSCN^{2+}]_{eq}$ allows you to determine the concentrations of the other two ions at equilibrium. For each mole of $FeSCN^{2+}$ ions produced, one less mole of Fe^{3+} ions will be found in the solution (see the 1:1 ratio of coefficients in the equation on the previous page). The $[Fe^{3+}]$ can be determined by:

$$[Fe^{3+}]_{eq} = [Fe^{3+}]_i - [FeSCN^{2+}]_{eq}$$

Because one mole of SCN^- is used up for each mole of $FeSCN^{2+}$ ions produced, $[SCN^-]_{eq}$ can be determined by:

$$[SCN^-]_{eq} = [SCN^-]_i - [FeSCN^{2+}]_{eq}$$

Knowing the values of $[Fe^{3+}]_{eq}$, $[SCN^-]_{eq}$, and $[FeSCN^{2+}]_{eq}$, you can now calculate the value of K_c, the equilibrium constant.

OBJECTIVE

In this experiment, you will determine the equilibrium constant, K_c, for the following chemical reaction:

$$Fe^{3+}(aq) + SCN^-(aq) \longleftrightarrow FeSCN^{2+}(aq)$$

iron(III)   thiocyanate   thiocyanoiron(III)

MATERIALS

computer	0.0020 M KSCN
Vernier computer interface*	0.0020 M Fe(NO$_3$)$_3$ (in 1.0 M HNO$_3$)
Logger*Pro*	0.200 M Fe(NO$_3$)$_3$ (in 1.0 M HNO$_3$)
Vernier Colorimeter or Spectrometer	four pipets
1 plastic cuvette	pipet bulb or pipet pump
five 20 × 150 mm test tubes	three 100 mL beakers
thermometer or Temperature Probe	tissues (preferably lint-free)

* No interface is required if using a Spectrometer

PROCEDURE

Both Colorimeter and Spectrometer Users

1. Obtain and wear goggles.

2. Label four 20 × 150 mm test tubes 1–4. Pour about 30 mL of 0.0020 M Fe(NO$_3$)$_3$ into a clean, dry 100 mL beaker. Pipet 5.0 mL of this solution into each of the four labeled test tubes. Use a pipet pump or bulb to pipet all solutions. **CAUTION:** *Fe(NO$_3$)$_3$ solutions in this experiment are prepared in 1.0 M HNO$_3$ and should be handled with care.* Pour about 25 mL of the 0.0020 M KSCN into another clean, dry 100 mL beaker. Pipet 2, 3, 4 and 5 mL of this solution into Test Tubes 1-4, respectively. Obtain about 25 mL of distilled water in a 100 mL beaker. Then pipet 3, 2, 1 and 0 mL of distilled water into Test Tubes 1–4, respectively, to bring the total volume of each test tube to 10 mL. Mix each solution *thoroughly* with a stirring rod. Be sure to clean and dry the stirring rod after each mixing. Measure and record the temperature of one of the above solutions to use as the temperature for the equilibrium constant, K_c. Volumes added to each test tube are summarized below:

Test Tube Number	Fe(NO$_3$)$_3$ (mL)	KSCN (mL)	H$_2$O (mL)
1	5	2	3
2	5	3	2
3	5	4	1
4	5	5	0

3. Prepare a standard solution of FeSCN^{2+} by pipetting 18 mL of 0.200 M Fe(NO$_3$)$_3$ into a 20 × 150 mm test tube labeled "5". Pipet 2 mL of 0.0020 M KSCN into the same test tube. Stir thoroughly.

4. Prepare a *blank* by filling a cuvette 3/4 full with distilled water. To correctly use cuvettes, remember:

   - Wipe the outside of each cuvette with a lint-free tissue.
   - Handle cuvettes only by the top edge of the ribbed sides.
   - Dislodge any bubbles by gently tapping the cuvette on a hard surface.
   - Always position the cuvette so the light passes through the clear sides.

Spectrometer Users Only (Colorimeter users proceed to the Colorimeter section)

5. Use a USB cable to connect the Spectrometer to the computer. Choose New from the File menu.

6. To calibrate the Spectrometer, place the blank cuvette into the cuvette slot of the Spectrometer, choose Calibrate ▶ Spectrometer from the Experiment menu. The calibration dialog box will display the message: "Waiting 90 seconds for lamp to warm up." After 90 seconds, the message will change to "Warmup complete." Click [OK].

7. Determine the optimal wavelength for creating the standard curve.

   a. Empty the water from the blank cuvette. Using the solution in Test Tube 1, rinse the cuvette twice with ~1 mL amounts and then fill it 3/4 full. Wipe the outside with a tissue, place it in the Spectrometer.
   b. Click [▶ Collect]. The absorbance *vs.* wavelength spectrum will be displayed. Click [■ Stop].
   c. To save your graph of absorbance *vs.* wavelength, select Store Latest Run from the Experiment menu.
   d. Click the Configure Spectrometer Data Collection icon, 🏠, on the toolbar. A dialog box will appear.
   e. Select Absorbance *vs.* Concentration under Set Collection Mode. The wavelength of maximum absorbance (λ max) is automatically identified. Click [OK].
   f. Proceed directly to Step 8.

Colorimeter Users Only

5. Connect the Colorimeter to the computer interface. Prepare the computer for data collection by opening the file "20 Equilibrium Constant" from the *Chemistry with Vernier* folder of Logger*Pro*.

6. Open the Colorimeter lid, insert the blank, and close the lid.

7. Calibrate the Colorimeter and prepare to test the standard solutions.

   a. Press the < or > button on the Colorimeter to select a wavelength of 470 nm (Blue).
   b. Press the CAL button until the red LED begins to flash and then release the CAL button.
   c. When the LED stops flashing, the calibration is complete.
   d. Empty the water from the blank cuvette. Use the solution in Test Tube 1 to rinse the cuvette twice with ~1 mL amounts and then fill it 3/4 full. Wipe the outside with a tissue, place it in the Colorimeter.

Both Colorimeter and Spectrometer Users

8. You are now ready to collect absorbance data for the four equilibrium systems and the standard solution.

    a. Leave the cuvette, containing the Test Tube 1 mixture, in the device (Spectrometer or Colorimeter). Close the lid on the Colorimeter. Click ▶ Collect . Click ⊕ Keep , type **1** (the trial number) in edit box, and click ☐ OK .

    b. Discard the cuvette contents as directed. Rinse the cuvette twice with the Test Tube 2 solution, fill the cuvette 3/4 full, and place it in the device. After the reading stabilizes, click ⊕ Keep , type **2** in the edit box, and click ☐ OK .

    c. Repeat the Step b procedure to find the absorbance of the solutions in Test Tubes 3, 4, and 5 (the standard solution).

    d. From the table, record the absorbance values for each of the five trials in your data table.

    e. Dispose of all solutions as directed by your instructor.

PROCESSING THE DATA

1. Write the K_c expression for the reaction in the Data and Calculation table.

2. Calculate the initial concentration of Fe^{3+}, based on the dilution that results from adding KSCN solution and water to the original 0.0020 M $Fe(NO_3)_3$ solution. See Step 2 of the procedure for the volume of each substance used in Trials 1-4. Calculate $[Fe^{3+}]_i$ using the equation (will be the same for all four test tubes):

$$[Fe^{3+}]_i = \frac{Fe(NO_3)_3 \text{ mL}}{\text{total mL}} \times (0.0020 \text{ M})$$

3. Calculate the initial concentration of SCN^-, based on its dilution by $Fe(NO_3)_3$ and water:

$$[SCN^-]_i = \frac{KSCN \text{ mL}}{\text{total mL}} \times (0.0020 \text{ M})$$

    In Test Tube 1, $[SCN^-]_i = (2 \text{ mL} / 10 \text{ mL})(0.0020 \text{ M}) = 0.00040 \text{ M}$. Calculate this for the other three test tubes.

4. $[FeSCN^{2+}]_{eq}$ is calculated using the formula:

$$[FeSCN^{2+}]_{eq} = \frac{A_{eq}}{A_{std}} \times [FeSCN^{2+}]_{std}$$

    where A_{eq} and A_{std} are the absorbance values for the equilibrium and standard test tubes, respectively, and $[FeSCN^{2+}]_{std} = (1/10)(0.0020) = 0.00020 \text{ M}$. Calculate $[FeSCN^{2+}]_{eq}$ for each of the four trials.

5. $[Fe^{3+}]_{eq}$: Calculate the concentration of Fe^{3+} at equilibrium for Trials 1-4 using the equation:

$$[Fe^{3+}]_{eq} = [Fe^{3+}]_i - [FeSCN^{2+}]_{eq}$$

6. $[SCN^-]_{eq}$: Calculate the concentration of SCN- at equilibrium for Trials 1-4 using the equation:

$$[SCN^-]_{eq} = [SCN^-]_i - [FeSCN^{2+}]_{eq}$$

7. Calculate K_c for Trials 1-4. Be sure to show the K_c expression and the values substituted in for each of these calculations.

8. Using your four calculated K_c values, determine an average value for K_c. How constant were your K_c values?

DATA AND CALCULATIONS

Absorbance	Trial 1	Trial 2	Trial 3	Trial 4
	_____	_____	_____	_____

Absorbance of standard (Trial 5)	_____	Temperature	_____ °C

K_c expression	$K_c =$			
$[Fe^{3+}]_i$				
$[SCN^-]_i$				
$[FeSCN^{2+}]_{eq}$				
$[Fe^{3+}]_{eq}$				
$[SCN^-]_{eq}$				
K_c value				

Average of K_c values

$K_c =$ _____ at _____ °C

Experiment
20

Chemical Equilibrium: Finding a Constant, K$_C$

1. The student pages with complete instructions for data-collection using LabQuest App, Logger *Pro* (computers), EasyData or DataMate (calculators), and DataPro (Palm handhelds) can be found on the CD that accompanies this book. See *Appendix A* for more information.

2. Preparation of solutions:

   0.0020 M potassium thiocyanate (0.39 g of solid KSCN per 2 L of solution) **HAZARD ALERT:** Moderately toxic by ingestion; emits toxic fumes of cyanide if strongly heated or in contact with concentrated acids. Hazard Code: C—Somewhat hazardous.

   0.0020 M iron (III) nitrate [0.81 g of solid Fe(NO$_3$)$_3$·9H$_2$O per 1 L (in 1.0 M HNO$_3$)] **HAZARD ALERT:** Fe(NO$_3$)$_3$·9H$_2$O: Strong oxidizer; skin and tissue irritant. Hazard Code: C—Somewhat hazardous.

   0.200 M iron (III) nitrate [8.08 g of solid Fe(NO$_3$)$_3$·9H$_2$O per 100 mL (in 1.0 M HNO$_3$)] **HAZARD ALERT:** Fe(NO$_3$)$_3$·9H$_2$O: Strong oxidizer; skin and tissue irritant. Hazard Code: C—Somewhat hazardous.

   1.0 M HNO$_3$ (61 mL of concentrated (16.4 M) HNO$_3$ per 1 L of solution) **HAZARD ALERT:** Corrosive; strong oxidant; toxic by inhalation; avoid contact with acetic acid and readily oxidized substances. Hazard Code: A—Extremely hazardous.

   The hazard information reference is: Flinn Scientific, Inc., *Chemical & Biological Catalog Reference Manual*, (800) 452-1261, www.flinnsci.com. See *Appendix D* of this book, *Chemistry with Vernier*, for more information.

3. It is important that the Fe(NO$_3$)$_3$ solutions be prepared in 1.0 M HNO$_3$ solution rather than water. The H$^+$ ion from the nitric acid prevents the formation of brown-colored complex ions, such as Fe(OH)$^{2+}$, that might interfere with colorimetric measurements. The preparation of equilibrium solutions in Test Tubes 1-4 *cannot* be done the day before making absorbance measurements with the Colorimeter. The colors of the equilibrium solutions fade significantly in 24 hours.

4. The four equilibrium constant values should be quite constant for a student lab team (see Sample Results on the next page). The equilibrium constant value may vary for different lab teams if different Colorimeters are used. Expect different lab teams using different Colorimeters to obtain an average K$_c$ value in the range of 130 to 170.

5. You should use a water-proof marker to make a reference mark on the top edge of one of the clear sides of all cuvettes. Students are reminded in the procedure to align this mark with the white reference mark to the right on the cuvette slot on the Colorimeter. The use of a single cuvette in the procedure is to eliminate errors introduced by slight variations in the absorbance of different plastic cuvettes. The two rinses done prior to adding a new solution can be accomplished very quickly.

6. As an alternative to having students prepare the standard FeSCN^{2+} solution, you can make up 100 mL of the standard ahead of time. Add 10.0 mL of 0.002 M KSCN to 90.0 mL of 0.200 M Fe(NO$_3$)$_3$. The absorbance of this solution affects all four of the equilibrium

constants; preparing larger quantities greatly reduces the chances for error and saves students time in completing the procedure.

7. There are two models of Vernier Colorimeters. The first model (rectangular shape) has three wavelength settings, and the newest model (a rounded shape) has four wavelength settings. The 470 nm wavelength of either model is used in this experiment. The newer model is an auto-ID sensor and supports automatic calibration (pressing the CAL button on the Colorimeter with a blank cuvette in the slot). If you have an older model Colorimeter, see www.vernier.com/til/1665.html for calibration information.

SAMPLE RESULTS

Absorbance	Trial 1	Trial 2	Trial 3	Trial 4
	0.153	0.214	0.277	0.345

Absorbance of standard (Trial 6)	Temperature
0.549	22.0°C

K_c expression	$K_c = \dfrac{[\text{FeSCN}^{2+}]}{[\text{Fe}^{3+}][\text{SCN}^-]}$			
$[\text{Fe}^{3+}]_i$	$\dfrac{5}{10} \times 0.0010$ M	$\dfrac{5}{10} \times (.0020)$ $= 0.0010$ M	$\dfrac{5}{10} \times (.0020)$ $= 0.0010$ M	$\dfrac{5}{10} \times (.0020)$ $= 0.0010$ M
$[\text{SCN}^-]_i$	$\dfrac{2}{10} \times (.0020)$ $= 0.00040$ M	$\dfrac{3}{10} \times (.0020)$ $= 0.00060$ M	$\dfrac{4}{10} \times (.0020)$ $= 0.00080$ M	$\dfrac{5}{10} \times (.0020)$ $= 0.00100$ M
$[\text{FeSCN}^{2+}]_{eq}$	$\dfrac{.153}{.549} \times (.0002)$ $= 5.57 \times 10^{-5}$ M	$\dfrac{.214}{.549} \times (.0002)$ $= 7.80 \times 10^{-5}$ M	$\dfrac{.277}{.549} \times (.0002)$ $= 1.01 \times 10^{-4}$ M	$\dfrac{.345}{.549} \times (.0002)$ $= 1.26 \times 10^{-4}$ M
$[\text{Fe}^{3+}]_{eq}$	$0.001 - 5.57 \times 10^{-5}$ $= 9.44 \times 10^{-4}$ M	$0.001 - 7.80 \times 10^{-5}$ $= 9.22 \times 10^{-4}$ M	$0.001 - 1.01 \times 10^{-4}$ $= 8.99 \times 10^{-4}$ M	$0.001 - 1.26 \times 10^{-4}$ $= 8.74 \times 10^{-4}$ M
$[\text{SCN}^-]_{eq}$	$0.0004 - 5.57 \times 10^{-5}$ $= 3.44 \times 10^{-4}$ M	$0.0006 - 7.80 \times 10^{-5}$ $= 5.22 \times 10^{-4}$ M	$0.0008 - 1.01 \times 10^{-4}$ $= 6.99 \times 10^{-4}$ M	$0.0010 - 1.26 \times 10^{-4}$ $= 8.74 \times 10^{-4}$ M
K_c value	172	162	161	165
Average of K_c values	$\dfrac{172 + 162 + 161 + 165}{5} =$ $K_c = 165$ at 22.0°C			

Computer

21

Household Acids and Bases

Many common household solutions contain acids and bases. Acid-base indicators, such as litmus and red cabbage juice, turn different colors in acidic and basic solutions. They can, therefore, be used to show if a solution is acidic or basic. An acid turns blue litmus paper red, and a base turns red litmus paper blue. The acidity of a solution can be expressed using the pH scale. Acidic solutions have pH values less than 7, basic solutions have pH values greater than 7, and neutral solutions have a pH value equal to 7.

In this experiment, you will use litmus and a computer-interfaced pH Sensor to determine the pH values of household substances. After adding red cabbage juice to the same substances, you will determine the different red cabbage juice indicator colors over the entire pH range.

OBJECTIVES

In this experiment, you will

- Use litmus paper and a pH Sensor to determine the pH values of household substances.
- Add cabbage juice to the same substances and determine different red cabbage juice indicator colors over the entire pH range.

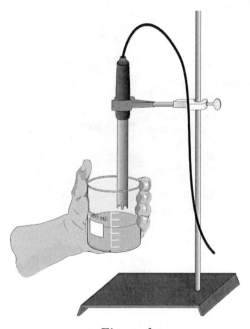

Figure 1

Chemistry with Vernier

21 - 1

MATERIALS

computer	household solutions
Vernier computer interface	7 small test tubes
Logger *Pro*	test-tube rack
Vernier pH Sensor	red and blue litmus paper
wash bottle	paper towel
distilled water	stirring rod
ring stand	red cabbage juice
utility clamp	250 mL beaker
sensor soaking solution	

PROCEDURE

1. Obtain and wear goggles. **CAUTION:** *Do not eat or drink in the laboratory.*

Part I Litmus Tests

2. Label 7 test tubes with the numbers 1–7 and place them in a test-tube rack.

3. Measure 3 mL of vinegar into test tube 1. Refer to the data table and fill each of the test tubes 2–7 to about the same level with its respective solution. **CAUTION:** *Ammonia solution is toxic. Its liquid and vapor are extremely irritating, especially to eyes. Drain cleaner solution is corrosive. Handle these solutions with care. Do not allow the solutions to contact your skin or clothing. Wear goggles at all times. Notify your teacher immediately in the event of an accident.*

4. Use a stirring rod to transfer one drop of vinegar to a small piece of blue litmus paper on a paper towel. Transfer one drop to a piece of red litmus paper on a paper towel. Record the results. Clean and dry the stirring rod each time.

5. Test solutions 2–7 using the same procedure. Be sure to clean and dry the stirring rod each time.

Part II Red Cabbage Juice Indicator

6. *After* you have finished the Part I litmus tests, add 3 mL of red cabbage juice indicator to each of the 7 test tubes. Record your observations. Dispose of the test-tube contents as directed by your teacher.

Part III pH Tests

7. Connect the pH Sensor to the computer interface. Prepare the computer to monitor pH by opening the file "21 Household Acids" from the *Chemistry with Vernier* folder.

8. Raise the pH Sensor from the sensor storage solution and set the solution aside. Use a wash bottle filled with distilled water to thoroughly rinse the tip of the sensor as demonstrated by your instructor. Catch the rinse water in a 250 mL beaker.

9. Get one of the 7 solutions in the small container supplied by your sensor. Raise the solution to the pH Sensor and swirl the solution about the sensor. When the pH reading stabilizes, record the pH value in your data table.

10. Prepare the pH Sensor for reuse.

   a. Rinse it with distilled water from a wash bottle.
   b. Place the sensor into the sensor soaking solution and swirl the solution about the sensor briefly.
   c. Rinse with distilled water again.

11. Determine the pH of the other solutions using the Step 9 procedure. You must clean the sensor, using the Step 10 procedure, between tests. When you are done, rinse the tip of the sensor with distilled water and return it to the sensor soaking solution.

DATA TABLE

Test Tube	Solution	Blue Litmus	Red Litmus	Red Cabbage Juice	pH
1	vinegar				
2	ammonia				
3	lemon juice				
4	soft drink				
5	drain cleaner				
6	detergent				
7	baking soda				

PROCESSING THE DATA

1. Which of the household solutions tested are acids? How can you tell?

2. Which of the solutions are bases? How can you tell?

3. What color(s) is red cabbage juice indicator in acids? In bases?

4. Can red cabbage juice indicator be used to determine the strength of acids and bases? Explain.

5. List advantages and disadvantages of litmus and red cabbage juice indicators.

Household Acids and Bases

1. The student pages with complete instructions for data-collection using LabQuest App, Logger *Pro* (computers), EasyData or DataMate (calculators), and DataPro (Palm handhelds) can be found on the CD that accompanies this book. See *Appendix A* for more information.

2. Red cabbage juice solution can be prepared by boiling red cabbage in water. Alternatively, you can grind up red cabbage in a blender, and then strain it with a sieve to obtain the juice.

3. The use of 13×100 mm test tubes is suggested.

4. Suggested solution concentrations:

   0.1 M ammonia (6.7 mL concentrated NH_3 per 1 L) **HAZARD ALERT:** Both liquid and vapor are extremely irritating—especially to eyes. Dispense in a hood and be sure an eyewash is accessible. Toxic by ingestion and inhalation. Serious respiratory hazard. Hazard Code: A—Extremely hazardous.

   0.1 M NaOH (0.40 g of NaOH per 100 mL to use as a drain cleaner solution) **HAZARD ALERT:** Corrosive solid; skin burns are possible; much heat evolves when added to water; very dangerous to eyes; wear face and eye protection when using this substance. Wear gloves. Hazard Code: B—Hazardous.

   1% detergent (1 g of solid detergent per 100 mL)

   1% baking soda (1 g of $NaHCO_3$ per 100 mL)

   Note: For safety reasons, 0.1 M NaOH solution is substituted for a drain cleaner (Drano) solution.

   The hazard information reference is: Flinn Scientific, Inc., *Chemical & Biological Catalog Reference Manual*, (800) 452-1261, www.flinnsci.com. See *Appendix D* of this book, *Chemistry with Vernier*, for more information.

5. The soda pop should be colorless.

6. To reduce the amount of litmus used, cut strips into 1 cm pieces.

7. We suggest that you have the pH Sensor connected to the computer for your students.

8. Be sure to demonstrate the pH Sensor cleaning sequence as you pre-lab the experiment.

9. Solutions can be made available to the students in 50 or 100 mL beakers. The 3 mL portions in Step 3 can be poured directly from the beakers, and pH tests can be made directly in the beakers.

10. The computer procedure has students record pH values from the live readouts (without clicking on ▶ Collect). Another possibility is to have students use the Selected Events mode for each of the 7 trials. In Logger*Pro,* the file is already set up for this option. Simply have your students click ▶ Collect, then click ⊛ Keep when the pH reading is stable. This saves the pH reading along with its trial number.

11. The electrode solution used in this experiment is pH 7 buffer solution. It can be purchased from chemical supply companies. Vernier Software & Technology sells a package of capsules for preparing buffer solutions of pH 4, 7, and 10 (Order Code PHB). The electrode solution can also be prepared using a recipe found in the *Handbook of Chemistry and Physics*. One such recipe specifies ingredients mixed in the ratio of 50 mL of 0.1 M KH_2PO_4 to 29.1 mL of 0.1 M NaOH. The 0.1 M KH_2PO_4 requires 13.7 g per 1 L of solution. The 0.1 M NaOH requires 4.0 g per 1 L of solution.

12. The sensor soaking solution can be contained in baby food jars, or some similar sealable containers. After your students finish using the pH Sensors, you should return the probes to the small plastic, buffer-filled bottles supplied by Vernier Software. The sensor soaking solution can be sealed and stored in its container for future use.

13. The stored pH calibration works well for this experiment.

SAMPLE RESULTS

Test Tube	Solution	Blue Litmus	Red Litmus	Red Cabbage Juice	pH
1	vinegar	pink	pink	red-pink	2.5
2	ammonia	blue	blue	blue-green	8.3
3	lemon juice	pink	pink	pink	2.7
4	soft drink	pink	pink	blue-pink	3.5
5	drain cleaner	blue	blue	yellow-green	12.0
6	detergent	blue	blue	blue-green	9.4
7	baking soda	blue	blue	blue	8.2

ANSWERS TO QUESTIONS

1. Vinegar, lemon juice, and soda pop are acidic. These solutions turn blue litmus pink, turn red cabbage juice pink, and they have pH values less than 7.

2. Ammonia, lye, detergent, and baking soda are basic. These solutions turn red litmus blue. They turn red cabbage juice blue, green, or yellow-green. They have pH values greater than 7.

3. Red cabbage juice indicator is pink in acids (red-pink in strong acids to blue-pink in weak acids). It is green in basic solutions (blue-green in weak bases to yellow-green in strong bases).

4. Yes. Red cabbage juice indicator can be used to determine the strengths of acids and bases. See the answer to Questions 2 and 3 above.

5. Although litmus has a long shelf life and no offensive odor, it only turns pink or blue. Although red cabbage juice turns a large range of colors, it has an unpleasant odor and it has a short shelf life.

Acid Rain

In this experiment, you will observe the formation of four acids that occur in acid rain:

- carbonic acid, H_2CO_3
- nitrous acid, HNO_2
- nitric acid, HNO_3
- sulfurous acid, H_2SO_3

Carbonic acid occurs when carbon dioxide gas dissolves in rain droplets of unpolluted air:

$$(1) \quad CO_2(g) + H_2O(l) \longrightarrow H_2CO_3(aq)$$

Nitrous acid and nitric acid result from a common air pollutant, nitrogen dioxide (NO_2). Most nitrogen dioxide in our atmosphere is produced from automobile exhaust. Nitrogen dioxide gas dissolves in rain drops and forms nitrous and nitric acid:

$$(2) \quad 2\ NO_2(g) + H_2O(l) \longrightarrow HNO_2(aq) + HNO_3(aq)$$

Sulfurous acid is produced from another air pollutant, sulfur dioxide (SO_2). Most sulfur dioxide gas in the atmosphere results from burning coal containing sulfur impurities. Sulfur dioxide dissolves in rain drops and forms sulfurous acid:

$$(3) \quad SO_2(g) + H_2O(l) \longrightarrow H_2SO_3(aq)$$

In the procedure outlined below, you will first produce these three gases. You will then bubble the gases through water, producing the acids found in acid rain. The acidity of the water will be monitored with a pH Sensor.

OBJECTIVES

In this experiment, you will

- Generate three gaseous oxides, CO_2, SO_2, and NO_2.
- Simulate the formation of acid rain by bubbling each of the three gases into water and producing three acidic solutions.
- Measure the pH of the three resulting acidic solutions to compare their relative strengths.

MATERIALS

computer	solid $NaNO_2$
Vernier computer interface	solid $NaHCO_3$
Logger*Pro*	solid $NaHSO_3$
Vernier pH Sensor	1 Beral pipet with 1.0 M HCl
wash bottle with distilled water	3 Beral pipets with a 2 cm stem
ring stand	3 Beral pipets with a 15 cm stem
100 mL beaker	utility clamp
20 × 150 mm test tube	tap water

PROCEDURE

1. Obtain and wear goggles.

2. Obtain three short-stem and three long-stem Beral pipets. Label the short-stem pipets with the formula of the solid they will contain: "NaHCO₃", "NaNO₂", and "NaHSO₃". Label the long-stem pipets with the formula of the gas they will contain: "CO₂", "NO₂" and "SO₂". You can use a 100 mL beaker to support the pipets.

3. Obtain a beaker containing solid NaHCO₃. Squeeze the bulb of the pipet labeled "NaHCO₃" to expel the air, and place the open end of the pipet into the solid NaHCO₃. When you release the bulb, solid NaHCO₃ will be drawn up into the pipet. Continue to draw solid into the pipet until there is enough to fill the curved end of the bulb, as shown in Figure 1.

Figure 1

4. Repeat the Step 3 procedure to add solid NaNO₂ and NaHSO₃ to the other two Beral pipets. **CAUTION:** *Avoid inhaling dust from these solids.*

5. Obtain a Beral pipet with 1.0 M HCl from your teacher. **CAUTION:** *HCl is a strong acid. Gently hold the pipet with the stem pointing up, so that HCl drops do not escape.* Insert the narrow stem of the HCl pipet into the larger opening of the pipet containing the solid NaHCO₃, as shown in Figure 2. Gently squeeze the HCl pipet to add about 20 drops of HCl solution to the solid NaHCO₃. When finished, remove the HCl pipet. Gently swirl the pipet that contains NaHCO₃ and HCl. Carbon dioxide, CO₂, is generated in this pipet. Place it in the 100 mL beaker, with the stem up, to prevent spillage.

6. Repeat the procedure in Step 5 by adding HCl to the pipet containing solid NaHSO₃. Sulfur dioxide, SO₂, is generated in this pipet.

Figure 2

7. Repeat the procedure in Step 5 by adding HCl to the pipet containing solid NaNO₂. Nitrogen dioxide, NO₂, is generated in this pipet. When you have finished this step, return the HCl pipet to your teacher. Leave the three gas-generating pipets in the 100 mL beaker until Step 10.

8. Use a utility clamp to attach a 20 × 150 mm test tube to the ring stand. Add about 4 mL of tap water to the test tube. Remove the pH Sensor from the pH storage solution, rinse it off with distilled water, and place it into the tap water in the test tube.

9. Connect the pH Sensor to the computer interface. Prepare the computer for data collection by opening the file "22 Acid Rain" from the *Chemistry with Vernier* folder. Check to see that the pH is between 6 and 8 for the water.

10. Squeeze all of the air from the bulb of the long-stem pipet labeled "CO₂". Keep the bulb completely collapsed and insert the long stem of the pipet down into the gas-generating pipet labeled "NaHCO₃", as shown in Figure 3. Be sure the tip of the long-stem pipet remains above the liquid in the gas-generating pipet. Release the pressure on the bulb so that it draws gas up into it. Store the long-stem pipet and the gas-generating pipet in the 100 mL beaker.

11. Repeat the procedure in Step 10 using the pipets labeled "NaNO₂" and "NO₂".

12. Repeat the procedure in Step 10 using the pipets labeled "NaHSO₃" and "SO₂".

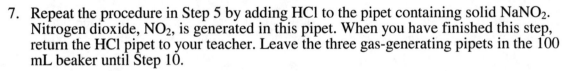
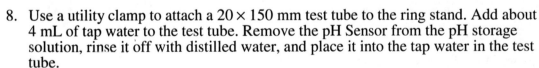
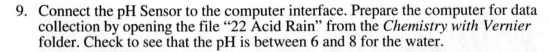

Figure 3

13. Insert the long-stem pipet labeled "CO_2" into the test tube, alongside the pH Sensor, so that its tip extends into the water to the bottom of the test tube (see Figure 4).

14. To begin collecting data, click ▶Collect. After 15 seconds have elapsed, gently squeeze the bulb of the pipet so that bubbles of CO_2 *slowly* bubble up through the solution. Use both hands to squeeze *all* of the gas from the bulb. When data collection stops after 120 seconds, examine the data in the table and determine the initial pH value (before CO_2 was added) and the final pH value (after CO_2 was added and pH stabilized). To confirm these two values, click the Statistics button, [STAT], and examine the minimum and maximum values in the pH box displayed on the graph. Record the initial and final pH values in your data table. Close the Statistics box by clicking in the upper left corner of the box.

15. Rinse the tip of the pH Sensor thoroughly with distilled water and return it to the sensor storage solution. Discard the contents of the test tube as directed by your teacher. Rinse the test tube *thoroughly* with tap water. Add 4 mL of tap water to the test tube. Place the pH Sensor in the test tube and check to see that the input display shows a pH value that is about the same as the previous initial pH. If not, rinse the test tube again.

16. From the Experiment menu, choose Store Latest Run. This stores the data so it can be used later, but it will be still be displayed while you do your second and third trials.

17. Repeat Steps 13–16 using NO_2 gas.

18. Repeat Steps 13–14 using SO_2 gas. When you are finished, rinse the pH Sensor with distilled water and return it to the sensor storage solution. Discard the six pipets as directed by your Instructor.

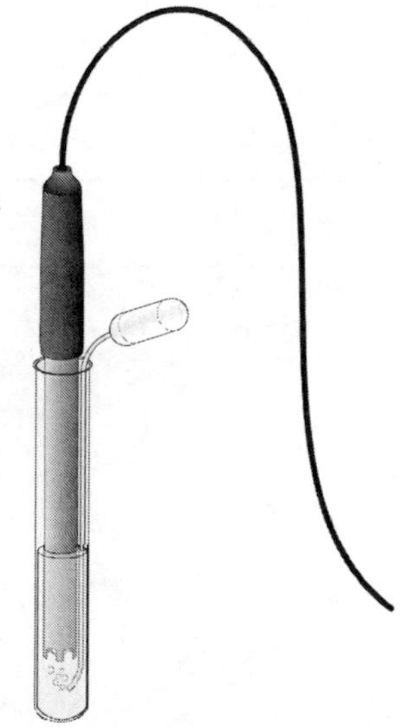

Figure 4

19. Label all three curves by choosing Text Annotation from the Insert menu, and typing "carbon dioxide" (or "nitrogen dioxide", or "sulfur dioxide") in the edit box.

20. Print copies of the graph, with all three data sets displayed.

PROCESSING THE DATA

1. For each of the three gases, calculate the change in pH (ΔpH), by subtracting the initial pH from the final pH. Record these values in your data table.

2. In this experiment, which gas caused the smallest drop in pH?

3. Which gas (or gases) caused the largest drop in pH?

4. Coal from western states such as Montana and Wyoming is known to have a lower percentage of sulfur impurities than coal found in the eastern United States. How would burning low-sulfur coal lower the level of acidity in rainfall? Use specific information about gases and acids to answer the question.

5. High temperatures in the automobile engine cause nitrogen and oxygen gases from the air to combine to form nitrogen oxides. What two acids in acid rain result from the nitrogen oxides in automobile exhaust?

6. Which gas and resulting acid in this experiment would cause rainfall in *unpolluted* air to have a pH value less than 7 (sometimes as low as 5.6)?

7. Why would acidity levels usually be lower (pH higher) in actual rainfall than the acidity levels you observed in this experiment? Rainfall in the United States generally has a pH between 4.5 and 6.0.

DATA AND CALCULATIONS TABLE

Gas	Initial pH	Final pH	Change in pH (ΔpH)
CO_2			
NO_2			
SO_2			

TEACHER INFORMATION

Acid Rain

1. The student pages with complete instructions for data-collection using LabQuest App, Logger *Pro* (computers), EasyData or DataMate (calculators), and DataPro (Palm handhelds) can be found on the CD that accompanies this book. See *Appendix A* for more information.

2. The 1.0 M HCl solution can be prepared by adding 8.6 mL of concentrated acid per 100 mL of solution. **HAZARD ALERT:** Highly toxic by ingestion or inhalation; severely corrosive to skin and eyes. Hazard Code: A—Extremely hazardous.

   Draw the HCl solution into the Beral pipets through the short, narrow stem. Since a trial requires approximately 1 mL of 1.0 M HCl, or a total of 3 mL for 3 gases, fill the bulb 3/4 full (3–4 mL).

3. Solid $NaHCO_3$, $NaHSO_3$, and $NaNO_2$ can be placed in 100 mL beakers to a depth of 1–2 cm.

   HAZARD ALERTS:

   Sodium bisulfite: Severe irritant to skin and tissue as an aqueous solution; moderately toxic. Hazard Code: C—Somewhat hazardous.

   Sodium nitrite: Strong oxidizer; fire and explosion risk if heated; highly toxic by ingestion and inhalation. Hazard Code: B—Hazardous.

   The hazard information reference is: Flinn Scientific, Inc., *Chemical & Biological Catalog Reference Manual*, (800) 452-1261, www.flinnsci.com. See *Appendix D* of this book, *Chemistry with Vernier*, for more information.

4. Thin-stem Beral pipets may be purchased from Flinn Scientific. The part numbers are:

AP1718	Pkg. of 20
AP1444	Pkg. of 500

   At a price of about $0.05 each, the pipets may be considered disposable. You can empty the pipets and discard or recycle them after using. You may also choose to empty, clean, and reuse the pipets.

5. One advantage of the microscale version of this experiment is that it avoids the odors of the two noxious gases, NO_2 and SO_2. Very little of either gas escapes into the room. You can operate the laboratory ventilation system during the experiment as a further precaution.

6. To make a narrower stem for the gas-collecting pipets and HCl pipets, it is necessary to stretch out the stem of the Beral pipet. To do this, place the pipet bulb in the palm of one hand with your thumb against the stem where it joins the bulb. Firmly grip the middle of the stem with your other hand and pull hard on the stem until it yields to the pressure and stretches out to a uniform narrow diameter. You can easily stretch it to the length needed for the gas-collecting pipets. Cut off the stems to a length of 15 cm for the gas pipets, and to a length of 4 cm for the HCl pipets. For the gas-generating pipet, cut the stem of a new Beral pipet to a length of 2 cm. Since it has a wider stem, the HCl and gas-collecting pipets will easily fit into it.

7. The directions in the experiment call for the use of a 100 mL beaker as a support for the Beral pipets. The pipets are placed in the beaker in an upright position, with the bulbs down. Test tube racks for 13×100 mm test tubes or 24 well microscale well plates also work well as supports for the Beral pipets.

8. The procedure directs students to obtain the 1.0 M HCl from the teacher, and then return it. For safety reasons, we felt it might be better for teachers to directly account for pipets containing HCl. The directions also have students return the used gas-generating and gas-collecting pipets at the end of the period. Whether you choose to dispose, recycle, or reuse the pipets, we recommend that your students not empty or clean the pipets. This way, accidents that might result from carelessly squeezing pipets containing HCl can be avoided. Empty the gas-generating pipets under a fume hood.

9. If you choose to reuse the gas-collecting pipets, you need to ensure that they are perfectly dry. The SO_2 and NO_2 gases are highly soluble, even in small droplets of water. Draw air in and out of the pipets 10 to 15 times to dry the bulbs.

10. To save time, you may choose to perform Step 3 of the procedure ahead of time. Students have very little difficulty adding the $NaHCO_3$ and $NaHSO_3$ powders to the Beral pipets, but have more trouble adding the larger granules of $NaNO_2$.

11. The equations for the production of each of the gases, as performed in this experiment, are:

    Carbon dioxide: $NaHCO_3(s) + HCl(aq) \longrightarrow NaCl(aq) + H_2O(1) + CO_2(g)$

    Sulfur dioxide: $NaHSO_3(s) + HCl(aq) \longrightarrow NaCl(aq) + H_2O(1) + SO_2(g)$

    Nitrogen dioxide: $3\ NaNO_2(s) + 3\ HCl(aq) \longrightarrow 3\ NaCl(aq) + HNO_3(aq) + 2\ NO(g) + H_2O$

    $\qquad\qquad\qquad 2\ NO(g) + O_2(g) \longrightarrow 2\ NO_2(g)$

12. Even though the procedure calls for tap water, distilled water can also be used. We use tap water because it normally contains enough dissolved CO_2, HCO_3^-, and CO_3^{2-} to give it a small amount of buffering capacity. This stabilizes the pH reading when the pH Sensor is first placed in the water and avoids fluctuations or gradual changes in pH that students sometimes encounter with distilled water. In the sample graphs on the next page, the buffering effect causes a smaller drop in pH in the first 5–10 seconds after the gas is added, followed by a more rapid drop.

13. This is a good time to discuss the topic of anhydrides with your students. All three of these gases are oxides of non-metals and represent good examples of *acidic anhydrides*.

14. A 20×150 mm test tube works well in this experiment. Test tubes size 18×150 mm will not easily allow the narrow stem of the pipet to fit alongside the pH Sensor.

15. The stored pH calibration works well for this experiment.

SAMPLE RESULTS

Gas	Initial pH	Final pH	Change in pH (ΔpH)
CO_2	7.15	5.93	−1.22
NO_2	7.10	2.74	−4.36
SO_2	7.06	2.54	−4.52

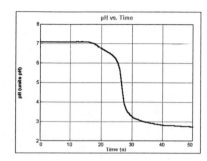 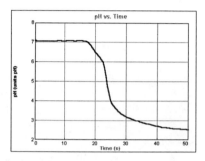

NO_2 Dissolving in Water   *SO_2 Dissolving in Water*

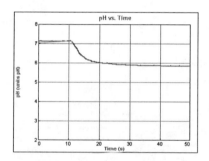

CO_2 Dissolving in Water

ANSWERS TO QUESTIONS

2. Carbon dioxide, CO_2, caused the smallest drop in pH ($\Delta pH = -1.22$).

3. Sulfur dioxide, SO_2, caused the largest drop in pH ($\Delta pH = -4.52$). Nitrogen dioxide, NO_2, causes a drop in pH about the same as SO_2 ($\Delta pH = -4.36$).

4. When low-sulfur coal is burned, it produces less sulfur dioxide. With lower concentrations of sulfur dioxide in the atmosphere, less sulfurous acid will be produced by the reaction:

$$SO_2(g) + H_2O(1) \longrightarrow H_2SO_3(aq)$$

5. Nitrous acid, HNO_2, and nitric acid, HNO_3, are produced by the reaction:

$$2\,NO_2(g) + H_2O(g) \longrightarrow HNO_2(aq) + HNO_3(aq)$$

6. Carbon dioxide gas, a natural component of the atmosphere, dissolves in rainwater and forms carbonic acid, H_2CO_3.

7. The acidity level is lower in actual rainfall because the concentration of SO_2, NO_2, and CO_2 gases in the atmosphere is much lower than in this experiment.

Titration Curves of Strong and Weak Acids and Bases

In this experiment you will react the following combinations of strong and weak acids and bases:

- Hydrochloric acid, HCl (strong acid), with sodium hydroxide, NaOH (strong base)
- Hydrochloric acid, HCl (strong acid), with ammonia, NH_3 (weak base)
- Acetic acid, $HC_2H_3O_2$ (weak acid), with sodium hydroxide, NaOH (strong base)
- Acetic acid, $HC_2H_3O_2$ (weak acid), with ammonia, NH_3 (weak base)

A computer-interfaced pH Sensor will be placed in one of the acid solutions. A solution of one of the bases will slowly drip from a buret into the acid solution at a constant rate. As base is added to the acid, you should see a gradual change in pH until the solution gets close to the equivalence point. At the equivalence point, equal numbers of moles of acid and base have been added. Near the equivalence point, a rapid change in pH occurs. Beyond the equivalence point, where more base has been added than acid, you should again observe more gradual changes in pH. A titration curve is normally a plot of pH versus *volume* of titrant. In this experiment, however, we will monitor and plot pH versus *time*, and assume that time is proportional to volume of base. The volume being delivered by the buret per unit time should be nearly constant.

One objective of this lab is to observe differences in shapes of titration curves when various strengths of acids and bases are combined. You will also learn about the function and selection of appropriate acid-base indicators in this experiment. In order to do several other experiments in this lab manual, you need to be able to interpret the shape of a titration curve.

OBJECTIVES

In this experiment, you will

- Observe differences in shapes of titration curves when various strengths of acids and bases are combined.
- Learn about the function and selection of appropriate acid-base indicators.
- Learn how to interpret the shape of a titration curve.

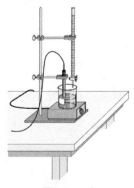

Figure 1

MATERIALS

computer	0.10 M NaOH
Vernier computer interface	0.10 M NH$_3$
Logger*Pro*	0.10 M HCl
Vernier pH Sensor	0.10 M HC$_2$H$_3$O$_2$
magnetic stirrer (if available)	50 mL buret
stirring bar	ring stand
250 mL beaker	2 utility clamps
phenolphthalein indicator	distilled water
wash bottle	

PROCEDURE

1. Obtain and wear goggles.

2. Place 8 mL of 0.1 M HCl solution into a 250 mL beaker. Add about 100 mL of distilled water. Add 3 drops of phenolphthalein acid-base indicator. **CAUTION:** *Handle the hydrochloric acid with care. It can cause painful burns if it comes in contact with the skin.*

3. Place the beaker onto a magnetic stirrer and add a small stirring bar. Turn on the stirrer and adjust it to a slow stirring speed.

4. Use a utility clamp to suspend a pH Sensor on a ring stand as shown in Figure 1. Situate the pH Sensor in the HCl solution and adjust its position toward the side of the beaker so that it is not struck by the stirring bar.

5. Obtain a 50 mL buret and rinse the buret with a few mL of the 0.1 M NaOH solution. Fill the buret to about the 0 mL mark. **CAUTION:** *Sodium hydroxide solution is caustic. Avoid spilling it on your skin or clothing.*

6. Connect the pH Sensor to the computer interface. Prepare the computer for data collection by opening the file "23 Titration Curves" from the *Chemistry with Vernier* folder. The pH reading should be between 2.0 and 3.0 for the HCl solution.

7. You are now ready to begin monitoring data. Click ▶ Collect . Carefully open the buret stopcock to provide a dripping rate of about 1 drop per second. Do not worry if the rate is somewhat faster or slower when you first start; initial additions of base will have very little effect on the pH.

8. Watch to see if the phenolphthalein changes color before, at the same time, or after the rapid change in pH at the equivalence point. **Note:** Time is being displayed in the table. If phenolphthalein is a suitable indicator for this reaction, it should change from clear to red at about the same time as the jump in pH occurs. In your data table, record the time when the phenolphthalein indicator changes color. When data collection has ended after 250 seconds, turn the buret stopcock to stop the flow of NaOH titrant.

9. Label the point on your graph where the indicator changes color.

   a. Choose Text Annotation from the Insert menu.

   b. Click in the text box and type "Color change".

   c. Click and drag the arrow head to the appropriate place on the graph.

   d. If you want, you can also move the text box. The cursor will become a hand when at the edge of the text box. At this point you can grab and move the box.

10. To print a graph of pH *vs.* time:

    a. Click on the graph. In the title edit box that appears, type the trial and acid and base strength. Click [OK].

    b. Print copies of the graph.

11. You can read pH and time values along the pH curve by clicking the Examine button, ⚡. As you move the mouse cursor across the graph, pH and time data points are displayed in the examine box on the graph. Determine the approximate time for the equivalence point; that is, for the biggest jump in pH in the steep vertical region of the curve. Record this time in the data table. Rinse the pH Sensor and return it to the sensor storage solution. Dispose of the beaker contents as directed by your teacher. Clean and dry the 250 mL beaker for the next trial. **Note:** You do not need to save or store your data for any of the four trials.

12. Repeat the procedure using NaOH titrant and acetic acid solution, $HC_2H_3O_2$. **CAUTION:** *Handle the solutions with care.* You do not need to refill the buret. Add 8 mL of 0.10 M $HC_2H_3O_2$ solution to the 250 mL beaker. Add about 100 mL of distilled water and 3 drops of phenolphthalein to the beaker. Rinse the tip of the sensor and position it in the acid solution as you did in Step 4. Repeat Steps 7-11 of the procedure.

13. Repeat the procedure using NH_3 titrant and HCl solution. **CAUTION:** *Handle the solutions with care.* Drain the remaining NaOH from the buret and dispose of it as directed by your teacher. Rinse the 50 mL buret with a few mL of the 0.1 M NH_3 solution. Fill the buret with NH_3 to about the 0 mL mark. Add 8 mL of 0.10 M HCl solution to the 250 mL beaker. Add about 100 mL of distilled water and 3 drops of phenolphthalein to the beaker. Rinse the sensor and position it in the acid solution as you did in Step 4. Repeat Steps 7-11 of the procedure.

14. Repeat the procedure using NH_3 titrant and $HC_2H_3O_2$ solution. **CAUTION:** *Handle the solutions with care.* You do not need to refill the buret. Add 8 mL of 0.10 M $HC_2H_3O_2$ solution to the 250 mL beaker. Add about 100 mL of distilled water and 3 drops of phenolphthalein to the beaker. Rinse the sensor and position it in the acid solution as you did in Step 4. Repeat Steps 7–11 of the procedure.

PROCESSING THE DATA

1. Examine the time data for each of the Trials 1–4. In which trial(s) did the indicator change color at about the same time as the large increase in pH occurred at the equivalence point? In which trial(s) was there a significant difference in these two times?

2. Phenolphthalein changes from clear to red at a pH value of about 9. According to your results, with which combination(s) of strong or weak acids and bases can phenolphthalein be used to determine the equivalence point?

3. On each of the four printed graphs, draw a horizontal line from a pH value of 9 on the vertical axis to its intersection with the titration curve. In which trial(s) does this line intersect the nearly vertical region of the curve? In which trial(s) does this line miss the nearly vertical region of the curve?

4. Compare your answers to Questions 1 and 3. By examining a titration curve, how can you decide which acid-base indicator to use to find the equivalence point?

5. Methyl red is an acid-base indicator that changes color at a pH value of about 5. From what you learned in this lab, methyl red could be used to determine the equivalence point of what combination of strong or weak acids and bases?

6. Of the four titration curves, which combination of strong or weak acids and bases had the longest vertical region of the equivalence point? The shortest?

7. The acid-base reaction between HCl and NaOH produces a solution with a pH of 7 at the equivalence point (NaCl + H_2O). Why does an acid-base indicator that changes color at pH 5 or 9 work just as well for this reaction as one that changes color at pH 7?

8. In general, how does the shape of a curve with a weak specie (NH_3 or $HC_2H_3O_2$) differ from the shape of a curve with a strong specie (NaOH or HCl)?

9. Complete each of the equations in the table.

DATA TABLE

Trial	Equation for acid-base reaction	Time of indicator color change	Time at equivalence point
1	NaOH + HCl $\longrightarrow$	s	s
2	NaOH + $HC_2H_3O_2$ $\longrightarrow$	s	s
3	NH_3 + HCl $\longrightarrow$	s	s
4	NH_3 + $HC_2H_3O_2$ $\longrightarrow$	s	s

Titration Curves of Weak and Strong Acids and Bases

1. The student pages with complete instructions for data-collection using LabQuest App, Logger *Pro* (computers), EasyData or DataMate (calculators), and DataPro (Palm handhelds) can be found on the CD that accompanies this book. See *Appendix A* for more information.

2. Preparation of solutions:

   0.1 M HCl (8.6 mL of concentrated HCl per 1 L of solution) **HAZARD ALERT:** Highly toxic by ingestion or inhalation; severely corrosive to skin and eyes. Hazard Code: A—Extremely hazardous.

   0.1 M $HC_2H_3O_2$ (5.7 mL of concentrated acetic acid, $HC_2H_3O_2$, per 1 L solution) **HAZARD ALERT:** Corrosive to skin and tissue; moderate fire risk (flash point: 39°C); moderately toxic by ingestion and inhalation. Hazard Code: A—Extremely hazardous.

   0.1 M NH_3 (6.7 mL of concentrated NH_3 per 1 L of solution) **HAZARD ALERT:** Both liquid and vapor are extremely irritating—especially to eyes. Dispense in a hood and be sure an eyewash is accessible. Toxic by ingestion and inhalation. Serious respiratory hazard. Hazard Code: A—Extremely hazardous.

   0.1 M NaOH (4.0 g of solid NaOH per 1 L of solution) **HAZARD ALERT:** Corrosive solid; skin burns are possible; much heat evolves when added to water; very dangerous to eyes; wear face and eye protection when using this substance. Wear gloves. Hazard Code: B—Hazardous.

   The hazard information reference is: Flinn Scientific, Inc., *Chemical & Biological Catalog Reference Manual*, (800) 452-1261, www.flinnsci.com. See *Appendix D* of this book, *Chemistry with Vernier*, for more information.

3. If drops of NaOH or NH_3 reach the sensor in the beaker too quickly (before they have been diluted in the acid solution), they may cause a spike in the curve. This can be avoided by dripping the base into the center of the whorl created by the stirring bar. The pH Sensor should be positioned near the side of the beaker.

4. The stored pH calibration works well for this experiment.

5. Other acid-base indicators may be used in this experiment. We use phenolphthalein because it is used in other experiments in this book. When using more than one indicator, consider having different lab groups use different indicators, with each group using one indicator for all four trials.

6. To reduce the number of printings, students can choose to Store Latest Run to store their first run data as Run 1—then, they can collect their second set of data. The data from both runs will then be displayed in the data table (as Run 1 and Latest). In a like manner, they may continue to save their third and fourth runs. Prior to printing, they can display as many of these runs on one graph as they wish. **Note:** DataMate and EasyData users are limited to a total of three runs.

7. A concentration of 0.1 M was chosen for safety reasons and to reduce odors from the ammonia and acetic acid solutions.

8. This lab is designed to be completed in a 50 minute class period. If you have less time available, you may want to assign only two acid-base combinations per student group. Time can be saved if the same base is used for both trials, so there is no need to refill the buret.

9. You may want to discuss with your students the effect of the response time of the pH Sensor. In a normal acid-base titration, the pH Sensor has enough time (5-10 seconds) for its pH reading to stabilize. In this experiment, the pH Sensor has only 1 second to respond before another drop is added. Near the equivalence point, the curve may not be as steep as in actual titrations. This may result in a 1-3 second difference between the time you see the indicator change color and the large jump in pH. **Note:** Vernier pH Sensors manufactured after January 1, 2006 have a faster response, so this will be less of a problem.

10. This experiment can also be used as a teacher demonstration. Follow the same procedure outlined in the experiment.

ANSWERS TO QUESTIONS

1. In Trials 1 and 2, the indicator color changed at about the same time as the large increase in pH. In Trials 3 and 4, the large increase in pH occurred before any change in indicator color.

2. Phenolphthalein can be used to determine the equivalence point in the reaction between a strong acid and strong base, or a weak acid and a strong base.

3. In Trials 1 and 2, the horizontal line intersects the nearly vertical region of the curve. In Trials 3 and 4, the horizontal line intersects the curve to the right of the vertical region or misses it altogether.

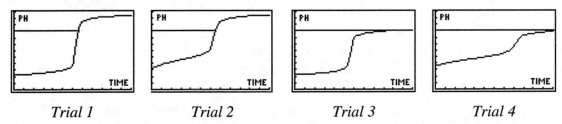

Trial 1   *Trial 2*   *Trial 3*   *Trial 4*

4. If the pH value at which the indicator changes color intersects the nearly vertical region of the curve, the indicator is suitable for that reaction.

5. Methyl red would be a suitable indicator for the reaction between a strong acid and a strong base, or a strong acid and a weak base. A pH 5 value falls within the nearly vertical region of these curves.

6. The longest vertical section resulted from a strong acid reacting with a strong base. The shortest vertical section resulted from a weak acid reacting with a weak base.

7. The very long vertical region of the curve at the equivalence point allows for a wide range of suitable indicators. Whether the indicator changes color at pH 5, 7, or 9, approximately the same volume of NaOH would be needed to reach the equivalence point.

8. A weak specie results in a continuous change in pH before or after the equivalence point. This effect shortens the length of the nearly vertical region of the curve at lower or higher pH values. A strong specie results in a very slight changes in pH before or after the equivalence point. This effect lengthens the nearly vertical region of the titration curve.

9. See the sample data table below.

SAMPLE RESULTS

Trial	Equation for acid-base reaction	Time of indicator color change	Time at equivalence point
1	$NaOH + HCl \longrightarrow NaCl + H_2O$	130 s	127 s
2	$NaOH + HC_2H_3O_2 \longrightarrow NaC_2H_3O_2 + H_2O$	128 s	127 s
3	$NH_3 + HCl \longrightarrow NH_4Cl$	186 s	114 s
4	$NH_3 + HC_2H_3O_2 \longrightarrow NH_4C_2H_3O_2$	130 s	130 s

SAMPLE RESULTS

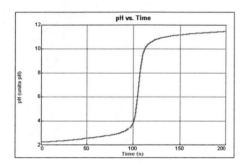

NaOH and HCl

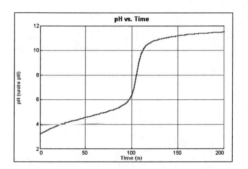

NaOH and HC$_3$H$_3$O$_2$

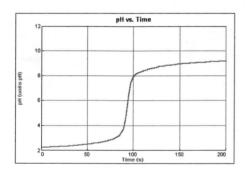

NH$_3$ and HCl

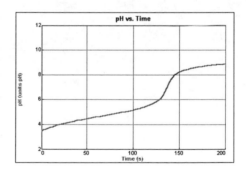

NH$_3$ and HC$_2$H$_3$O$_2$

Acid-Base Titration

A titration is a process used to determine the volume of a solution needed to react with a given amount of another substance. In this experiment, you will titrate hydrochloric acid solution, HCl, with a basic sodium hydroxide solution, NaOH. The concentration of the NaOH solution is given and you will determine the unknown concentration of the HCl. Hydrogen ions from the HCl react with hydroxide ions from the NaOH in a one-to-one ratio to produce water in the overall reaction:

$$H^+(aq) + Cl^-(aq) + Na^+(aq) + OH^-(aq) \longrightarrow H_2O(l) + Na^+(aq) + Cl^-(aq)$$

When an HCl solution is titrated with an NaOH solution, the pH of the acidic solution is initially low. As base is added, the change in pH is quite gradual until close to the equivalence point, when equimolar amounts of acid and base have been mixed. Near the equivalence point, the pH increases very rapidly, as shown in Figure 1. The change in pH then becomes more gradual again, before leveling off with the addition of excess base.

In this experiment, you will use a computer to monitor pH as you titrate. The region of most rapid pH change will then be used to determine the equivalence point. The volume of NaOH titrant used at the equivalence point will be used to determine the molarity of the HCl.

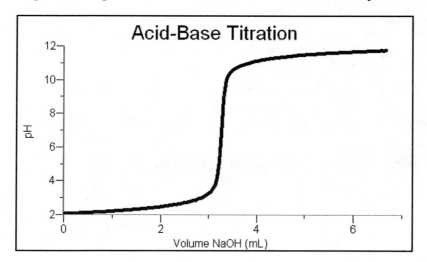

Figure 1

OBJECTIVES

In this experiment, you will

- Use a pH Sensor to monitor changes in pH as sodium hydroxide solution is added to a hydrochloric acid solution.
- Plot a graph of pH *vs.* volume of sodium hydroxide solution added.
- Use the graph to determine the equivalence point of the titration.
- Use the results to calculate the concentration of the hydrochloric acid solution.

MATERIALS

Materials for *both* Method 1 (buret) *and* Method 2 (Drop Counter)

computer
Vernier computer interface
Logger*Pro*
Vernier pH Sensor
HCl solution, unknown concentration
~0.1 M NaOH solution
pipet bulb or pump

magnetic stirrer (if available)
stirring bar or Microstirrer (if available)
wash bottle
distilled water
ring stand
1 utility clamp
250 mL beaker

Materials required *only* for Method 1 (buret)

50 mL buret
10 mL pipet

2nd utility clamp
2nd 250 mL beaker

Materials required *only* for Method 2 (Drop Counter)

Vernier Drop Counter
60 mL reagent reservoir
5 mL pipet or graduated 10 mL pipet

100 mL beaker
10 mL graduated cylinder

CHOOSING A METHOD

Method 1 has the student deliver volumes of NaOH titrant from a buret. After titrant is added, and pH values have stabilized, the student is prompted to enter the buret reading manually and a pH-volume data pair is stored.

Method 2 uses a Vernier Drop Counter to take volume readings. NaOH titrant is delivered drop by drop from the reagent reservoir through the Drop Counter slot. After the drop reacts with the reagent in the beaker, the volume of the drop is calculated, and a pH-volume data pair is stored.

METHOD 1: Measuring Volume Using a Buret

1. Obtain and wear goggles.

2. Add 50 mL of distilled water to a 250 mL beaker. Use a pipet bulb (or pipet pump) to pipet 10.0 mL of the HCl solution into the distilled water in the 250 mL beaker. **CAUTION:** *Handle the hydrochloric acid with care. It can cause painful burns if it comes in contact with the skin.*

3. Place the beaker on a magnetic stirrer and add a stirring bar. If no magnetic stirrer is available, you need to stir with a stirring rod during the titration.

4. Use a utility clamp to suspend a pH Sensor on a ring stand as shown here. Position the pH Sensor in the HCl solution and adjust its position so it will not be struck by the stirring bar. Turn on the magnetic stirrer, and adjust it to a medium stirring rate (with no splashing of solution).

5. Obtain approximately 60 mL of ~0.1 M NaOH solution in a 250 mL beaker. Obtain a 50 mL buret and rinse the buret with a few mL of the ~0.1 M NaOH solution. Use a utility clamp to

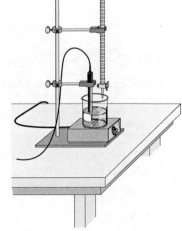

attach the buret to the ring stand as shown here. Fill the buret a little above the 0.00 mL level of the buret with ~0.1 M NaOH solution. Drain a small amount of NaOH solution into the beaker so it fills the buret tip *and* leaves the NaOH at the 0.00 mL level of the buret. Record the precise concentration of the NaOH solution in your data table. Dispose of the waste solution from this step as directed by your teacher. **CAUTION:** *Sodium hydroxide solution is caustic. Avoid spilling it on your skin or clothing.*

6. Connect the pH Sensor to the computer interface. Prepare the computer for data collection by opening the file "24a Acid-Base Titration" from the *Chemistry with Vernier* folder. Check to see that the pH value is between 2 and 3.

7. Before adding NaOH titrant, click ▶ Collect and monitor pH for 5-10 seconds. Once the displayed pH reading has stabilized, click ⊛ Keep. In the edit box, type "0" (for 0 mL added). Press the ENTER key or click ⌐ OK ⌐ to store the first data pair for this experiment.

8. You are now ready to begin the titration. This process goes faster if one person manipulates and reads the buret while another person operates the computer and enters volumes.

    a. Add the next increment of NaOH titrant (enough to raise the pH about 0.15 units). When the pH stabilizes, again click ⊛ Keep. In the edit box, type the current buret reading, to the nearest 0.01 mL. Press ENTER or click ⌐ OK ⌐. You have now saved the second data pair for the experiment.

    b. Continue adding NaOH solution in increments that raise the pH by about 0.15 units and enter the buret reading after each increment. Proceed in this manner until the pH is 3.5.

    c. When a pH value of approximately 3.5 is reached, change to a one-drop increment. Enter a new buret reading after each increment. Note: It is important that all increment volumes in this part of the titration be equal; that is, one-drop increments.

    d. After a pH value of approximately 10 is reached, again add larger increments that raise the pH by about 0.15 pH units, and enter the buret level after each increment.

    e. Continue adding NaOH solution until the pH value remains constant.

9. When you have finished collecting data, click ■ Stop. Dispose of the beaker contents as directed by your teacher.

10. Print copies of the table and the graph.

11. If time permits, repeat the procedure.

METHOD 2: Measuring Volume with a Drop Counter

1. Obtain and wear goggles.

2. Connect the pH Sensor to CH 1 of the computer interface. Lower the Drop Counter onto a ring stand and connect its cable to DIG/SONIC 1.

3. Add 40 mL of distilled water to a 100 mL beaker. Use a pipet bulb (or pipet pump) to pipet 5.00 mL of the HCl solution into the 100 mL beaker with distilled water. **CAUTION:** *Handle the hydrochloric acid with care. It can cause painful burns if it comes in contact with the skin.*

4. Obtain approximately 40 mL of ~0.1 M NaOH solution in a 250 mL beaker. Record the precise NaOH concentration in your data table. **CAUTION:** *Sodium hydroxide solution is caustic. Avoid spilling it on your skin or clothing.*

5. Obtain the plastic 60 mL reagent reservoir. **Note:** The bottom valve will be used to open or close the reservoir, while the top valve will be used to finely adjust the flow rate. For now, close both valves by turning the handles to a horizontal position.

   Rinse it with a few mL of the ~0.1 M NaOH solution. Use a utility clamp to attach the reagent reservoir to the ring stand. Add the remainder of the NaOH solution to the reagent reservoir.

   Drain a small amount of NaOH solution into the 250 mL beaker so it fills the reservoir's tip. To do this, turn both valve handles to the vertical position for a moment, then turn them both back to horizontal.

6. Prepare the computer for data collection by opening the file "24b Acid-Base (Drop Count)" from the *Chemistry with Vernier* folder.

7. To calibrate drops so that a precise volume of titrant is recorded in units of milliliters:

   a. From the Experiment menu, choose Calibrate ▶ DIG 1: Drop Counter (mL).

   b. Proceed by one of these two methods:
      - If you have previously calibrated the drop size of your reagent reservoir and want to continue with the same drop size, select the Manual button, enter the number of Drops/mL, and click ⌷ OK ⌷. Then proceed directly to Step 8.
      - If you want to perform a new calibration, select the Automatic button, and continue with Step c below.

   c. Place a 10 mL graduated cylinder directly below the slot on the Drop Counter, lining it up with the tip of the reagent reservoir.

   d. Open the bottom valve on the reagent reservoir (vertical). Keep the top valve closed (horizontal).

   e. Click the Start button.

   f. Slowly open the top valve of the reagent reservoir so that drops are released at a slow rate (~1 drop every two seconds). You should see the drops being counted on the computer screen.

   g. When the volume of NaOH solution in the graduated cylinder is between 9 and 10 mL, close the bottom valve of the reagent reservoir.

   h. Enter the precise volume of NaOH (read to the nearest 0.1 mL) in the edit box. Record the number of Drops/mL displayed on the screen for possible future use.

   i. Click ⌷ OK ⌷. Discard the NaOH solution in the graduated cylinder as indicated by your instructor and set the graduated cylinder aside.

8. Assemble the apparatus.

   a. Place the magnetic stirrer on the base of the ring stand.

   b. Insert the pH Sensor through the large hole in the Drop Counter.

   c. Attach the Microstirrer to the bottom of the pH Sensor, as shown in the small picture. Rotate the paddle wheel of the Microstirrer and make sure that it does not touch the bulb of the pH Sensor.

   d. Adjust the positions of the Drop Counter and reagent reservoir so they are both lined up with the center of the magnetic stirrer.

   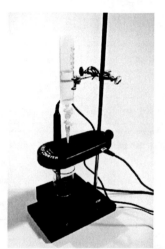

   e. Lift up the pH Sensor, and slide the beaker containing the HCl solution onto the magnetic stirrer. Lower the pH Sensor into the beaker. Check to see that the pH value is between 1.5 and 2.5.

   f. Adjust the position of the Drop Counter so that the Microstirrer on the pH Sensor is just touching the bottom of the beaker.

   g. Adjust the reagent reservoir so its tip is just above the Drop Counter slot.

9. Turn on the magnetic stirrer so that the Microstirrer is stirring at a fast rate.

10. You are now ready to begin collecting data. Click ▶ Collect. No data will be collected until the first drop goes through the Drop Counter slot. Fully open the bottom valve—the top valve should still be adjusted so drops are released at a rate of about 1 drop every 2 seconds. When the first drop passes through the Drop Counter slot, check the data table to see that the first data pair was recorded.

11. Continue watching your graph to see when a large increase in pH takes place—this will be the equivalence point of the reaction. When this jump in pH occurs, let the titration proceed for several more milliliters of titrant, then click ■ Stop. Turn the bottom valve of the reagent reservoir to a closed (horizontal) position.

12. Dispose of the beaker contents as directed by your teacher.

13. Print copies of the table.

14. Print copies of the graph.

15. If time permits, repeat the procedure.

PROCESSING THE DATA

1. Use your graph and data table to determine the volume of NaOH titrant used in each trial. Examine the data to find the largest increase in pH values upon the addition of 1 drop of NaOH solution. Find and record the NaOH volume just *before* and *after* this jump.

2. Determine the volume of NaOH added at the equivalence point. To do this, add the two NaOH values determined above and divide by two.

3. Calculate the number of moles of NaOH used.

4. Using the equation for the neutralization reaction given in the introduction, determine the number of moles of HCl used.

5. Calculate the HCl concentration using the volume of unknown HCl you pipeted out for each titration.

6. (Optional) If you did two titrations, determine the average [HCl] in mol/L.

DATA TABLE

Concentration of NaOH	M	M
NaOH volume added *before* the largest pH increase	mL	mL
NaOH volume added *after* the largest pH increase	mL	mL
Volume of NaOH added at equivalence point	mL	mL
Moles NaOH	mol	mol
Moles HCl	mol	mol
Concentration of HCl	mol/L	mol/L
Average [HCl]		M

EQUIVALENCE POINT DETERMINATION: An Alternate Method

An alternate way of determining the precise equivalence point of the titration is to take the first and second derivatives of the pH-volume data. The equivalence point volume corresponds to the *peak* (maximum) value of the first derivative plot, and to the volume where the second derivative equals *zero* on the second derivative plot.

1. In Method 1, view the first-derivative graph ($\Delta pH/\Delta vol$) by clicking the on the vertical-axis label (pH), and choose First Derivative. You may need to autoscale the new graph by clicking the Autoscale button, ▣.

   In Method 2, you can also view the first derivative graph (pH/Δvol) on Page 2 of the experiment file by clicking the Next Page button, ▣. On Page 2 you will see a plot of first derivative *vs.* volume.

2. In Method 1, view the second-derivative graph ($\Delta^2 pH/\Delta vol^2$) by clicking on the vertical-axis label, and choosing Second Derivative. In Method 2, view the second-derivative on Page 3 by clicking the Next Page button, ▣.

TEACHER INFORMATION

Acid-Base Titration

1. The student pages with complete instructions for data-collection using LabQuest App, Logger *Pro* (computers), EasyData or DataMate (calculators), and DataPro (Palm handhelds) can be found on the CD that accompanies this book. See *Appendix A* for more information.

2. The experiment "Titration Curves of Strong and Weak Acids and Bases" serves as a good introduction to this student experiment and can be done prior to this one.

3. There are two methods described in the student procedure. Method 1 has the student deliver volumes of NaOH titrant from a buret, so buret volumes must be read (and entered) manually. Method 2 uses a Vernier Drop Counter to take volume readings. Method 1 has students titrate 10 mL of HCl solution. Method 2 has students titrate 5 mL of HCl solution. The reason we use less HCl volume in Method 2 is to decrease the time required for each data collection. Sample data shown here are for Method 1.

4. The preparation of 0.1 M NaOH requires 4.0 g of NaOH per liter of solution. **HAZARD ALERT:** Corrosive solid; skin burns are possible; much heat evolves when added to water; very dangerous to eyes; wear face and eye protection when using this substance. Wear gloves. Hazard Code: B—Hazardous. Two options when indicating the concentration are:

   - Consider it to be 0.1000 M NaOH (the volume can also be read to four significant figures).
   - Standardize the solution with a standard acid solution and indicate its concentration to three or four significant figures (using a centigram or milligram balance to weigh out the acid).

5. Unknown samples with HCl concentrations in the 0.075 to 0.15 M range work well. The preparation of 0.075 M HCl requires 6.2 mL of concentrated HCl per liter of solution. HCl that is 0.15 M requires 12.5 mL of concentrated reagent per liter. **HAZARD ALERT:** Highly toxic by ingestion or inhalation; severely corrosive to skin and eyes. Hazard Code: A— Extremely hazardous.

   The hazard information reference is: Flinn Scientific, Inc., *Chemical & Biological Catalog Reference Manual*, (800) 452-1261, www.flinnsci.com. See *Appendix D* of this book, *Chemistry with Vernier*, for more information.

6. The Logger *Pro* experiment files for Experiment 24 include calculated columns for first and second derivatives of the pH-volume data. Directions are included in the student experiment (see the section, Equivalence Point Determination: An alternate Method). To help determine the equivalence-point volume, students can examine a graph of first derivative *vs.* volume ($\Delta pH/\Delta vol$) or second derivative *vs.* volume ($\Delta^2 pH/\Delta vol^2$). Sample first and second derivative plots are shown here using data from Method 1 of the experiment.

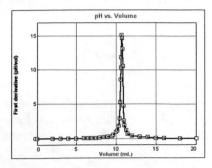
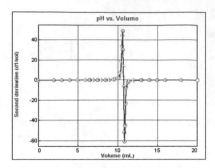

First derivative *Second derivative*

7. Consider having your students add two or three drops of phenolphthalein indicator at the beginning of each titration. They can then observe the phenolphthalein equivalence point and compare it with their pH equivalence point for each titration.

8. The stored pH calibration works well for this experiment. For more accurate pH readings, you (or your students) can do a 2-point calibration for each pH Sensor using pH-4 and pH-7 buffers.

9. If you are using a Vernier Electrode Amplifier (order code EA-BTA) with a detachable pH electrode and the Drop Counter, you will need use a special Logger *Pro* experiment file. The file is called "EA pH Drop Counter," and can be found in the Experiments folder, inside Probes & Sensors\Electrode Amplifier.

SAMPLE RESULTS

pH	Volume (mL)	pH	Volume (mL)
2.25	0.00	4.79	10.60
2.28	2.00	5.31	10.65
2.32	3.00	5.91	10.70
2.37	4.00	6.49	10.75
2.43	5.00	7.17	10.80
2.49	6.00	8.36	10.85
2.54	6.50	8.86	10.90
2.59	7.00	9.12	10.95
2.65	7.50	9.30	11.00
2.72	8.00	9.56	11.10
2.81	8.50	9.91	11.30
2.93	9.00	10.19	11.50
3.10	9.50	10.53	12.00
3.20	9.70	10.95	13.00
3.36	10.00	11.15	14.00
3.60	10.20	11.29	15.00
3.98	10.40	11.40	16.00
4.17	10.50	11.52	18.00
4.40	10.55	11.61	20.00

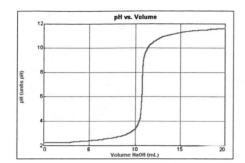 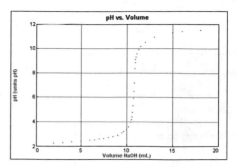

Acid-base titration for sodium hydroxide and hydrochloric acid

SAMPLE RESULTS

	Trial 1	Trial 2
Concentration of NaOH	0.1000 M	0.1000 M
NaOH volume added before largest pH increase	10.80 mL	10.90 mL
NaOH volume added after largest pH increase	10.85 mL	10.95 mL
Volume of NaOH added at equivalence point	$\dfrac{10.80 + 10.85}{2} = 10.825$ mL 10.83 mL	$\dfrac{10.90 + 10.95}{2} = 10.925$ mL 10.93 mL
Moles NaOH	(0.100 mol/L)(0.01083 L) = 0.00108 mol	(0.100 mol/L)(0.01093 L) = 0.00109 mol
Moles HCl	$\dfrac{1 \text{ mol HCl}}{1 \text{ mol NaOH}} \times 0.00108$ mol = 0.00108 mol	$\dfrac{1 \text{ mol HCl}}{1 \text{ mol NaOH}} \times 0.00109$ mol = 0.00109 mol
Concentration of HCl	$\dfrac{0.00108 \text{ mol}}{0.0100 \text{ L}} =$ 0.108 mol/L	$\dfrac{0.00109 \text{ mol}}{0.0100 \text{ L}} =$ 0.109 mol/L
Average [HCl]	$= \dfrac{0.108 + 0.109}{2} = 0.1085$ M 0.109 M	

Titration of a Diprotic Acid: Identifying an Unknown

A diprotic acid is an acid that yields two H^+ ions per acid molecule. Examples of diprotic acids are sulfuric acid, H_2SO_4, and carbonic acid, H_2CO_3. A diprotic acid dissociates in water in two stages:

(1) $H_2X(aq) \longleftrightarrow H^+(aq) + HX^-(aq)$

(2) $HX^-(aq) \longleftrightarrow H^+(aq) + X^{2-}(aq)$

Because of the successive dissociations, titration curves of diprotic acids have two equivalence points, as shown in Figure 1. The equations for the acid-base reactions occurring between a diprotic acid, H_2X, and sodium hydroxide base, NaOH, are

from the beginning to the first equivalence point:

(3) $H_2X + NaOH \longleftrightarrow NaHX + H_2O$

from the first to the second equivalence point:

(4) $NaHX + NaOH \longleftrightarrow Na_2X + H_2O$

from the beginning of the reaction through the second equivalence point (net reaction):

(5) $H_2X + 2\ NaOH \longleftrightarrow Na_2X + 2\ H_2O$

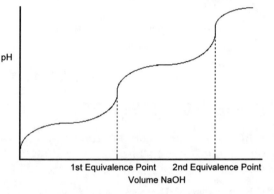

Figure 1

At the first equivalence point, all H^+ ions from the first dissociation have reacted with NaOH base. At the second equivalence point, all H^+ ions from *both* reactions have reacted (twice as many as at the first equivalence point). Therefore, the volume of NaOH added at the second equivalence point is exactly twice that of the first equivalence point (see Equations 3 and 5).

The primary purpose of this experiment is to identify an unknown diprotic acid by finding its molecular weight. A diprotic acid is titrated with NaOH solution of known concentration. Molecular weight (or molar mass) is found in g/mole of the diprotic acid. Weighing the original sample of acid will tell you its mass in grams. Moles can be determined from the volume of NaOH titrant needed to reach the first equivalence point. The volume and the concentration of NaOH titrant are used to calculate moles of NaOH. Moles of unknown acid equal moles of NaOH at the first equivalence point (see Equation 3). Once *grams* and *moles* of the diprotic acid are known, molecular weight can be calculated, in g/mole. Molecular weight determination is a common way of identifying an unknown substance in chemistry.

You may use either the first or second equivalence point to calculate molecular weight. The first is somewhat easier, because moles of NaOH are equal to moles of H_2X (see Equation 3). If the second equivalence point is more clearly defined on the titration curve, however, simply divide its NaOH volume by 2 to confirm the first equivalence point; or from Equation 5, use the ratio:

$$1\ mol\ H_2X\ /\ 2\ mol\ NaOH$$

OBJECTIVE

In this experiment, you will identify an unknown diprotic acid by finding its molecular weight.

MATERIALS

Materials for *both* Method 1 (buret) *and* Method 2 (Drop Counter)

computer	magnetic stirrer
Vernier computer interface	stirring bar or Vernier Microstirrer
Logger*Pro*	wash bottle
Vernier pH Sensor	distilled water
unknown diprotic acid, 0.120 g	ring stand
~0.1 M NaOH solution (standardized)	1 utility clamp
milligram balance	250 mL beaker

Materials required *only* for Method 1 (buret)

50 mL buret	2nd utility clamp
2nd 250 mL beaker	

Materials required *only* for Method 2 (Drop Counter)

Vernier Drop Counter	100 mL beaker
60 mL reagent reservoir	10 mL graduated cylinder

CHOOSING A METHOD

Method 1 has the student deliver volumes of NaOH titrant from a buret. After titrant is added, and pH values have stabilized, the student is prompted to enter the buret reading manually and a pH-volume data pair is stored.

Method 2 uses a Vernier Drop Counter to take volume readings. NaOH titrant is delivered drop by drop from the reagent reservoir through the Drop Counter slot. After the drop reacts with the reagent in the beaker, the volume of the drop is calculated, and a pH-volume data pair is stored.

METHOD 1: Measuring Volume Using a Buret

1. Obtain and wear goggles.

2. Weigh out about 0.120 g of the unknown diprotic acid on a piece of weighing paper. Record the mass to the nearest 0.001 g in your data table. Transfer the unknown acid to a 250 mL beaker and dissolve in 100 mL of distilled water. **CAUTION:** *Handle the solid acid and its solution with care. Acids can harm your eyes, skin, and respiratory tract.*

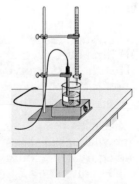

3. Place the beaker on a magnetic stirrer and add a stirring bar. If no magnetic stirrer is available, you need to stir with a stirring rod during the titration.

4. Use a utility clamp to suspend a pH Sensor on a ring stand as shown here. Position the pH Sensor in the diprotic acid solution and adjust its position toward the outside of the beaker so it will not be struck by the stirring bar. Turn on the magnetic stirrer, and adjust it to a medium stirring rate (with no splashing of solution).

5. Obtain approximately 60 mL of ~0.1 M NaOH solution in a 250 mL beaker. Obtain a 50 mL buret and rinse the buret with a few mL of the ~0.1 M NaOH solution. Record the precise concentration of the NaOH solution in your data table. Use a utility clamp to attach the buret to the ring stand. Fill the buret a little above the 0.00 mL level of the buret. Drain a small amount of NaOH solution into the beaker so it fills the buret tip *and* leaves the NaOH at the 0.00 mL level of the buret. Dispose of the waste solution from this step as directed by your teacher. **CAUTION:** *Sodium hydroxide solution is caustic. Avoid spilling it on your skin or clothing.*

6. Connect the pH Sensor to the computer interface. Prepare the computer for data collection by opening the file "25a Titration Dip Acid" from the *Chemistry with Vernier* folder of Logger*Pro*.

7. You are now ready to begin the titration. This process goes faster if one person manipulates and reads the buret while another person operates the computer and enters buret readings.

    a. Before adding NaOH titrant, click ▶ Collect and monitor pH for 5-10 seconds. Once the pH has stabilized, click ✪ Keep. In the edit box, type "0" (for 0 drops added), and press ENTER to store the first data pair for this experiment.

    b. Add enough NaOH to raise the pH by about 0.20 units. When the pH stabilizes, again click ✪ Keep. In the edit box, type the current buret reading, to the nearest 0.01 mL. Press ENTER. You have now saved the second data pair for the experiment.

    c. Continue adding NaOH solution in increments that raise the pH about 0.20 units and enter the buret reading after each addition. Proceed in this manner until the pH is 3.5.

    d. When pH 3.5 is reached, change to 2-drop increments. Enter the buret reading after each increment.

    e. After pH 4.5 is reached, again add larger increments that raise the pH by about 0.20 units and enter the buret reading after each addition. Continue in this manner until a pH of 7.5 is reached.

    f. When pH 7.5 is reached, change to 2-drop increments. Enter the buret reading after each increment.

    g. When pH 10 is reached, again add larger increments that raise the pH by 0.20 units. Enter the buret reading after each increment. Continue in this manner until you reach a pH of 11.

8. When you have finished collecting data, click ■ Stop. Dispose of the beaker contents as directed by your teacher.

9. Print a copy of the table. Then print a copy of the graph.

METHOD 2: Measuring Volume with a Drop Counter

1. Obtain and wear goggles.

2. Connect the pH Sensor to CH 1 of the computer interface. Lower the Drop Counter onto a ring stand and connect its cable to DIG/SONIC 1.

3. Weigh out about 0.120 g of the unknown diprotic acid on a piece of weighing paper. Record the mass to the nearest 0.001 g in your data table. Transfer the unknown acid to a 100 mL beaker and dissolve in 40 mL of distilled water. **CAUTION:** *Handle the solid acid and its solution with care. Acids can harm your eyes, skin, and respiratory tract.*

4. Obtain approximately 40 mL of ~0.1 M NaOH solution in a 250 mL beaker. Record the precise NaOH concentration in your data table. **CAUTION:** *Sodium hydroxide solution is caustic. Avoid spilling it on your skin or clothing.*

5. Obtain the plastic 60 mL reagent reservoir. **Note:** The bottom valve will be used to open or close the reservoir, while the top valve will be used to finely adjust the flow rate. For now, close both valves by turning the handles to a horizontal position.

   Rinse it with a few mL of the ~0.1 M NaOH solution. Use a utility clamp to attach the reagent reservoir to the ring stand. Add the remainder of the NaOH solution to the reagent reservoir.

   Drain a small amount of NaOH solution into the 250 mL beaker so it fills the reservoir's tip. To do this, turn both valve handles to the vertical position for a moment, then turn them both back to horizontal.

6. Prepare the computer for data collection by opening the file "25b Titration (Drop Count)" from the *Chemistry with Vernier* folder. Check to see that the pH value is between 1.5 and 2.5.

7. To calibrate drops so that a precise volume of titrant is recorded in units of milliliters:

   a. From the Experiment menu, choose Calibrate ▶ DIG 1: Drop Counter (mL).
   b. Proceed by one of these two methods:
      • If you have previously calibrated the drop size of your reagent reservoir and want to continue with the same drop size, select the Manual button, enter the number of Drops/mL, and click [OK]. Then proceed directly to Step 8.
      • If you want to perform a new calibration, select the Automatic button, and continue with Step c below.
   c. Place a 10 mL graduated cylinder directly below the slot on the Drop Counter, lining it up with the tip of the reagent reservoir.
   d. Open the bottom valve on the reagent reservoir (vertical). Keep the top valve closed (horizontal).
   e. Click the Start button.
   f. Slowly open the top valve of the reagent reservoir so that drops are released at a slow rate (~1 drop every two seconds). You should see the drops being counted on the computer screen.
   g. When the volume of NaOH solution in the graduated cylinder is between 9 and 10 mL, close the bottom valve of the reagent reservoir.
   h. Enter the precise volume of NaOH (read to the nearest 0.1 mL) in the edit box. Record the number of Drops/mL displayed on the screen for possible future use.
   i. Click [OK]. Discard the NaOH solution in the graduated cylinder as indicated by your instructor and set the graduated cylinder aside.

8. Assemble the apparatus.

   a. Place the magnetic stirrer on the base of the ring stand.
   b. Insert the pH Sensor through the large hole in the Drop Counter.
   c. Attach the Microstirrer to the bottom of the pH Sensor, as shown here. Rotate the paddle wheel of the Microstirrer and make sure that it does not touch the bulb of the pH Sensor.

   d. Adjust the positions of the Drop Counter and reagent
      reservoir so they are both lined up with the center of the
      magnetic stirrer.

   e. Lift up the pH Sensor, and slide the beaker containing the HCl
      solution onto the magnetic stirrer. Lower the pH Sensor into
      the beaker. Check to see that the pH value is between 1.5
      and 2.5.

   f. Adjust the position of the Drop Counter so that the
      Microstirrer on the pH Sensor is just touching the bottom of
      the beaker.

   g. Adjust the reagent reservoir so its tip is just above the Drop
      Counter slot.

9. Turn on the magnetic stirrer so that the Microstirrer is stirring at
   a fast rate.

10. You are now ready to begin collecting data. Click ▶ Collect. No data will be collected until the
    first drop goes through the Drop Counter slot. Fully open the bottom valve—the top valve
    should still be adjusted so drops are released at a rate of about 1 drop every 2 seconds. When
    the first drop passes through the Drop Counter slot, check the data table to see that the first
    data pair was recorded.

11. Continue watching your graph to see when a large increase in pH takes place—this will be
    the equivalence point of the reaction. Then, a second large increase occurs at the second
    equivalence point. When this jump in pH occurs, let the titration proceed for several more
    milliliters of titrant, then click ■ Stop. Turn the bottom valve of the reagent reservoir to a
    closed (horizontal) position.

12. Dispose of the beaker contents as directed by your teacher.

13. Print copies of the table. Then print copies of the graph.

EQUIVALENCE POINT DETERMINATION: An Alternate Method

An alternate way of determining the precise equivalence point of the titration is to take the first
and second derivatives of the pH-volume data. The equivalence point volume corresponds to the
peak (maximum) value of the first derivative plot, and to the volume where the second derivative
equals *zero* on the second derivative plot.

1. In Method 1, view the first-derivative graph ($\Delta pH/\Delta vol$) by clicking the on the vertical-axis
   label (pH), and choose First Derivative. You may need to autoscale the new graph by
   clicking the Autoscale button, Ⓐ.

   In Method 2, you can also view the first derivative graph (pH/Δvol) on Page 2 of the
   experiment file by clicking on the Next Page button, ▣. On Page 2 you will see a plot of first
   derivative *vs.* volume.

2. In Method 1, view the second-derivative graph ($\Delta^2 pH/\Delta vol^2$) by clicking on the vertical-axis
   label, and choosing Second Derivative. In Method 2, view the second-derivative on Page 3
   by clicking on the Next Page button, ▣.

PROCESSING THE DATA

1. On your printed graph, one of the two equivalence points is usually more clearly defined than the other; the two-drop increments near the equivalence points frequently result in larger increases in pH (a steeper slope) at one equivalence point than the other. Indicate the more clearly defined equivalence point (first or second) in your data table.

2. Use your graph and data table to determine the volume of NaOH titrant used for the equivalence point you selected in Step 1. To do so, examine the data to find the largest increase in pH values during the 2-drop additions of NaOH. Find the NaOH volume just *before* this jump. Then find the NaOH volume *after* the largest pH jump. Underline both of these data pairs on the printed data table and record them in your data table.

3. Determine the volume of NaOH added at the equivalence point you selected in Step 1. To do this, add the two NaOH volumes determined in Step 2, and divide by two. For example:

$$\frac{12.34 + 12.44}{2} = 12.39 \text{ mL}$$

4. Calculate the number of moles of NaOH used at the equivalence point you selected in Step 1.

5. Determine the number of moles of the diprotic acid, H_2X. Use Equation 3 or Equation 5 to obtain the ratio of moles of H_2X to moles of NaOH, depending on which equivalence point you selected in Step 1.

6. Using the mass of diprotic acid you measured out in Step 1 of the procedure, calculate the molecular weight of the diprotic acid, in g/mol.

7. From the following list of five diprotic acids, identify your unknown diprotic acid.

Diprotic Acid	Formula	Molecular weight
Oxalic Acid	$H_2C_2O_4$	90
Malonic Acid	$H_2C_3H_2O_4$	104
Maleic Acid	$H_2C_4H_2O_4$	116
Malic Acid	$H_2C_4H_4O_5$	134
Tartaric Acid	$H_2C_4H_4O_6$	150

8. Determine the percent error for your molecular weight value in Step 6.

9. For the *alternate* equivalence point (the one you did *not* use in Step 1), use your graph and data table to determine the volume of NaOH titrant used. Examine the data to find the largest increase in pH values during the 2-drop additions of NaOH. Find the NaOH volume just before and after this jump. Underline both of these data pairs on the printed data table and record them in the Data and Calculations table. Note: Dividing or multiplying the other equivalence point volume by two may help you confirm that you have selected the correct two data pairs in this step.

10. Determine the volume of NaOH added at the alternate equivalence point, using the same method you used in Step 3.

11. On your printed graph, clearly specify the position of the equivalence point volumes you determined in Steps 3 and 10, using dotted reference lines like those in Figure 1. Specify the NaOH volume of each equivalence point on the horizontal axis of the graph.

DATA TABLE

Mass of diprotic acid	g
Concentration of NaOH	M

1. Equivalence point (indicate which one you will use in the calculations below)	first equivalence point ____ or second equivalence point ____
2. NaOH volume added before and after the largest pH increase	_____ mL _____ mL
3. Volume of NaOH added at the equivalence point	mL
4. Moles of NaOH	mol
5. Moles of diprotic acid, H_2X	mol
6. Molecular weight of diprotic acid	g/mol
7. Name, formula, and accepted molecular weight of the diprotic acid	_____ _____ ____ g/mol
8. Percent error	%

9. Alternate equivalence point (indicate the one used in the calculations below)	first equivalence point ____ or second equivalence point ____
10. NaOH volume added before and after the largest pH increase	_____ mL _____ mL
11. Volume of NaOH added at the alternate equivalence point	mL

EXTENSION

Using a half-titration method, it is possible to determine the acid dissociation constants, K_{a1} and K_{a2}, for the two dissociations of the diprotic acid in this experiment. The K_a expressions for the first and second dissociations, from Equations 1 and 2, are:

$$K_{a1} = \frac{[H^+][HX^-]}{[H_2X]} \qquad\qquad K_{a2} = \frac{[H^+][X^{2-}]}{[HX^-]}$$

The first half-titration point occurs when *one-half* of the H^+ ions in the first dissociation have been titrated with NaOH, so that $[H_2X] = [HX^-]$. Similarly, the second half-titration point occurs when one-half of the H^+ ions in the second dissociation have been titrated with NaOH, so that $[HX^-] = [X^{2-}]$. Substituting $[H_2X]$ for $[HX^-]$ in the K_{a1} expression, and $[HX^-]$ for $[X^{2-}]$ in the K_{a2} expressions, the following are obtained:

$$K_{a1} = [H^+] \qquad\qquad K_{a2} = [H^+]$$

Taking the base-ten log of both sides of each equation,

$$\log K_{a1} = \log[H^+] \qquad\qquad \log K_{a2} = \log[H^+]$$

Thus, the pH value at the first half-titration volume, Point 1 in Figure 2, is equal to the pK_{a1} value. The first half-titration point volume can be found by dividing the first equivalence point volume by two.

Similarly, the pH value at the second titration point, is equal to the pK_{a2} value. The second half-titration volume (Point 2 in Figure 2) is midway between the first and second equivalence point volumes (1st EP and 2nd EP). Use the method described below to determine the K_{a1} and K_{a2} values for the diprotic acid you identified in this experiment.

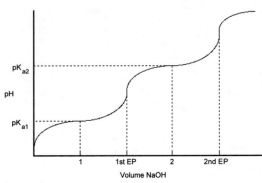

Figure 2

1. Determine the precise NaOH volume for the *first* half-titration point using one-half of the first equivalence point volume (determined in Step 2 or Step 9 of Processing the Data). Then determine the precise NaOH volume of the *second* half-titration point halfway between the first and second equivalence points.

2. On your graph of the titration curve, draw reference lines similar to those shown in Figure 2. Start with the first half-titration point volume (Point 1) and the second half-titration point volume (Point 2). Determine the pH values on the vertical axis that correspond to each of these volumes. Estimate these two pH values to the nearest 0.1 pH unit. These values are the pK_{a1} and pK_{a2} values, respectively. (Note: See if there are volume values in your data table similar to either of the half-titration volumes in Step 1. If so, use their pH values to confirm your estimates of pK_{a1} and pK_{a2} from the graph.)

3. From the pK_{a1} and pK_{a2} values you obtained in the previous step, calculate the K_{a1} and K_{a2} values for the two dissociations of the diprotic acid.

Titration of a Diprotic Acid: Identifying an Unknown

1. The student pages with complete instructions for data-collection using LabQuest App, Logger *Pro* (computers), EasyData or DataMate (calculators), and DataPro (Palm handhelds) can be found on the CD that accompanies this book. See *Appendix A* for more information.

2. The preparation of 0.1 M NaOH requires 4.00 g of NaOH per liter of solution. Here are two options for standardizing the solution, accurate to 3 or 4 significant figures.

    *   Standardize the NaOH solution using a standard acid solution, such as potassium hydrogen phthalate, $KHC_8H_4O_4$.
    *   Standardize the NaOH solution using maleic acid as your standard solution. Titrate the maleic acid to the second equivalence point. Follow the procedure outlined in this experiment. Instead of calculating the moles of maleic acid based on the concentration of NaOH, calculate the concentration of NaOH based on the moles of maleic acid.

3. The precision of the analytical balance you use is an important consideration in this experiment. The sample data was collected using a balance with a precision of 0.001 g. This results in three significant figures in the final molecular weight (117 g/mol). Using an analytical balance precise to 0.0001 g will gain a significant figure, since the standard acid and NaOH solutions can also be determined to four significant figures. Four significant figures can also be obtained for any NaOH volume over 10.00 mL.

4. We recommend the use of maleic acid as the unknown acid for all of your students. Guard the secret carefully! Other diprotic acids we tried give good resolution for only one of the equivalence points. Malonic acid provides satisfactory results.

    HAZARD ALERTS:

    Maleic acid: Moderately toxic by ingestion; severe body tissue irritant. Hazard Code: C—Somewhat hazardous.

    Malonic acid: Strong irritant; moderately toxic; when dissolved in water it is a strong acid; corrosive to eyes, skin and respiratory tract. Malonic acid has been used as a drug intermediate (precursor). This substance is regulated in some states like California. Check local regulations before purchasing. Hazard Code: B—Hazardous.

    Sodium Hydroxide: Corrosive solid; skin burns are possible; much heat evolves when added to water; very dangerous to eyes; wear face and eye protection when using this substance. Wear gloves. Hazard Code: B—Hazardous.

    The hazard information reference is: Flinn Scientific, Inc., *Chemical & Biological Catalog Reference Manual,* (800) 452-1261, www.flinnsci.com. See *Appendix D* of this book, *Chemistry with Vernier,* for more information.

5. There are two methods described in the student procedure. Method 1 has the student deliver volumes of NaOH titrant from a buret, so buret volumes must be read (and entered) manually. Method 2 uses a Vernier Drop Counter to take volume readings. Both methods have students weigh out 0.120 g of the diprotic acid, which provides 3 significant figures in calculations. Alert students that using this amount of acid requires about 15 minutes per trial

in Method 2. (You can cut this time in half using 0.060 g of diprotic acid, but you will also lose one significant figure.) Sample data shown on the following pages are for Method 1.

6. It is important to have a well-calibrated pH Sensor for this experiment, especially if you want your students to do the Extension exercise. Using the stored pH calibration in the software, check a pH Sensor in pH-7 buffer to see if it gives a reading close to pH of 7. If it does, then one option is to have your students use the stored calibration. For even better results use buffers of pH 4.0 and 7.0 to calibrate the pH Sensor. To do this:

First Calibration Point

a. Set up the data-collection software to calibrate the pH Sensor.
b. For the first calibration point, rinse the pH Sensor with distilled water, then place it into a buffer of pH 4.0.
c. Type **4** in the edit box as the pH value.
d. Swirl the sensor, wait until the voltage for stabilizes, then Keep the point.

Second Calibration Point

e. Rinse the pH Sensor with distilled water, and place it into a buffer of pH 7.0.
f. Type **7** in the edit box as the pH value for the second calibration point.
g. Swirl the sensor and wait until the voltage stabilizes, then Keep the point. Then select either OK or ☐ **Done** ☐ depending on the software. This completes the calibration.
h. You are now ready to collect data, using the calibration. Or, to use the calibration values for a specific pH Sensor at a later time, you can simply resave this experiment file—the new calibration is saved along with the experiment file itself.

7. As can be seen in the sample data above, the second equivalence point results in clearer resolution of the equivalence point volume. For this reason, students who accidentally overshoot the first equivalence point should be encouraged to continue the titration, rather than start over.

8. If you are using a Vernier Electrode Amplifier (order code EA-BTA) with a detachable pH electrode and the Drop Counter, you will need use a special Logger *Pro* experiment file. The file is called "EA pH Drop Counter," and can be found in the Experiments folder, inside Probes & Sensors\Electrode Amplifier.

9. If you choose to have your students do the Extension and determine the acid dissociation constants for maleic acid, you may want them to collect more data in the region of the half-titration points. As can be seen in the sample data, 0.20 pH-unit increments result in fairly large volume jumps in this part of the curve. Increments of about 0.5-1.0 mL will result in more reliable pH and pK_a data in these regions of the graph.

10. If students manually graph the data on high-quality graph paper, more precise values for Ka_1 and Ka_2 can be obtained.

11. The acid dissociation constant values are listed here for the possible unknown acids. You may provide your students with this list to help them in identifying the unknown acid, if they do the Extension exercise.

Diprotic Acid	K_{a1}	pK_{a1}	K_{a2}	pK_{a2}
Oxalic Acid	5.9×10^{-2}	1.23	6.4×10^{-5}	4.19
Malonic Acid	1.5×10^{-3}	2.83	2.0×10^{-6}	5.69
Maleic Acid	1.4×10^{-2}	1.83	8.6×10^{-7}	6.07
Malic Acid	3.9×10^{-4}	3.40	7.8×10^{-6}	5.11
Tartaric Acid	1.0×10^{-3}	2.98	4.6×10^{-6}	4.34

SAMPLE RESULTS

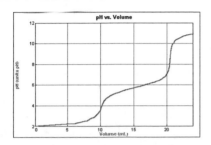
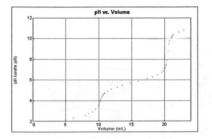

pH	Vol mL	pH	Vol mL	pH	Vol mL	pH	Vol mL
2.05	0.00	4.40	10.50	6.14	17.40	9.50	20.68
2.30	6.18	4.50	10.61	6.31	18.21	9.78	20.79
2.58	8.01	4.59	10.73	6.50	19.00	9.98	20.90
2.75	8.52	4.69	10.84	6.75	19.58	10.11	21.01
2.93	9.10	4.75	10.97	6.87	19.80	10.19	21.12
3.16	9.51	4.94	11.38	7.08	20.02	10.29	21.23
3.51	9.96	5.11	11.90	7.19	20.14	10.35	21.36
3.69	10.02	5.32	12.80	7.36	20.27	10.65	22.24
3.89	10.17	5.49	13.57	7.50	20.32	10.82	23.00
4.10	10.29	5.69	14.76	7.84	20.44	10.94	24.03
4.27	10.40	5.89	16.02	8.82	20.55		

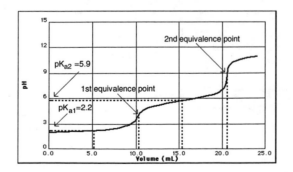

SAMPLE RESULTS

Mass of diprotic acid	0.119 g
Concentration of NaOH	0.0995 M

1. Equivalence point (indicate the one you will use you will use in the calculations below)	first equivalence point ____ or second equivalence point __X__
2. NaOH volume added before and after the largest pH increase	20.44 mL 20.55 mL
3. Volume of NaOH added at the equivalence point	$\dfrac{20.44 + 20.55}{2} =$ 20.50 mL
4. Moles of NaOH	(0.0995 mol/L)(0.02050 L) = 0.00204 mol
5. Moles of diprotic acid, H_2X	$(0.00204 \text{ mol NaOH}) \cdot \dfrac{1 \text{ mol } H_2X}{2 \text{ mol NaOH}} =$ 0.00102 mol
6. Molecular weight of diprotic acid	$\dfrac{0.119 \text{ g}}{0.00102 \text{ mol}} =$ 117 g/mol
7. Name, formula, and accepted molecular weight of the diprotic acid	maleic acid $H_2C_4H_2O_4$ 116 g/mol
8. Percent error	$\dfrac{|116 - 117|}{|116|} \cdot 100 =$ 0.9 %

9. Alternative equivalence point (indicate the one used below)	first equivalence point __X__ or second equivalence point ____
10. NaOH volume before and after largest pH increase (alternate equivalence point)	__10.17__ mL __10.29__ mL
11. Volume of NaOH added at the alternate equivalence point	$\dfrac{10.17 + 10.29}{2} =$ 10.23 mL

Using Conductivity to
Find an Equivalence Point

In this experiment, you will monitor conductivity during the reaction between sulfuric acid, H_2SO_4, and barium hydroxide, $Ba(OH)_2$, in order to determine the equivalence point. From this information, you can find the concentration of the $Ba(OH)_2$ solution. You will also see the effect of ions, precipitates, and water on conductivity. The equation for the reaction in this experiment is:

$$Ba^{2+}(aq) + 2\ OH^-(aq) + 2\ H^+(aq) + SO_4^{2-}(aq) \longrightarrow BaSO_4(s) + H_2O(l)$$

Before reacting, $Ba(OH)_2$ and H_2SO_4 are almost completely dissociated into their respective ions. Neither of the reaction products, however, is significantly dissociated. Barium sulfate is a precipitate and water is predominantly molecular.

As 0.02 M H_2SO_4 is slowly added to $Ba(OH)_2$ of unknown concentration, changes in the conductivity of the solution will be monitored using a Conductivity Probe. When the probe is placed in a solution that contains ions, and thus has the ability to conduct electricity, an electrical circuit is completed across the electrodes that are located on either side of the hole near the bottom of the probe body. This results in a conductivity value that can be read by the interface. The unit of conductivity used in this experiment is microsiemens per centimeter, or µS/cm.

Prior to doing the experiment, it is very important for you to hypothesize about the conductivity of the solution at various stages during the reaction. Would you expect the conductivity reading to be high or low, and increasing or decreasing, in each of these situations?

- When the Conductivity Probe is placed in $Ba(OH)_2$, prior to the addition of H_2SO_4.
- As H_2SO_4 is slowly added, producing $BaSO_4$ and H_2O.
- When the moles of H_2SO_4 added equal the moles of $BaSO_4$ originally present.
- As excess H_2SO_4 is added beyond the equivalence point.

OBJECTIVES

In this experiment, you will

- Hypothesize about the conductivity of a solution of sulfuric acid and barium hydroxide at various stages during the reaction.
- Use a Conductivity Probe to monitor conductivity during the reaction.
- See the effect of ions, precipitates, and water on conductivity.

CHOOSING A METHOD

Method 1 has the student deliver volumes of H_2SO_4 titrant from a buret. After titrant is added, and conductivity values have stabilized, the student is prompted to enter the buret reading manually and a conductivity-volume data pair is stored.

Method 2 uses a Vernier Drop Counter to take volume readings. H_2SO_4 titrant is delivered drop by drop from the reagent reservoir through the Drop Counter slot. After the drop reacts with the reagent in the beaker, the volume of the drop is calculated, and a conductivity-volume data pair is stored.

MATERIALS

Materials for *both* Method 1 (buret) *and* Method 2 (Drop Counter)

computer	magnetic stirrer (if available)
Vernier computer interface	stirring bar or Vernier Microstirrer
Logger*Pro*	100 mL graduated cylinder
Vernier Conductivity Probe	Phenolphthalein (optional)
60 mL of ~0.02 M H_2SO_4	ring stand
50 mL of $Ba(OH)_2$, unknown concentration	1 utility clamp
250 beaker	

Materials required *only* for Method 1 (buret)

50 mL buret	2nd utility clamp
2nd 250 mL beaker	

Materials required *only* for Method 2 (Drop Counter)

Vernier Drop Counter	100 mL beaker
60 mL reagent reservoir	10 mL graduated cylinder

METHOD 1: Measuring Volume Using a Buret

1. Obtain and wear goggles.

2. Measure out approximately 60 mL of ~0.02 M H_2SO_4 solution into a 250 mL beaker. Record the precise H_2SO_4 concentration in your data table. **CAUTION:** *H_2SO_4 is a strong acid, and should be handled with care.* Obtain a 50 mL buret and rinse the buret with a few mL of the H_2SO_4 solution. Use a utility clamp to attach the buret to the ring stand as shown here. Fill the buret a little above the 0.00 mL level of the buret. Drain a small amount of H_2SO_4 solution so it fills the buret tip *and* leaves the H_2SO_4 at the 0.00 mL level of the buret. Dispose of the waste solution from this step as directed by your instructor.

3. Measure out 50.0 mL of $Ba(OH)_2$ solution of unknown concentration using a 100 mL graduated cylinder. Transfer the solution to a clean, dry 250 mL beaker. Then add 120 mL of distilled water to the beaker. **CAUTION:** *$Ba(OH)_2$ is toxic. Handle it with care.*

4. Arrange the buret, Conductivity Probe, beaker containing $Ba(OH)_2$, and stirring bar as shown here. The Conductivity Probe should extend down into the $Ba(OH)_2$ solution to just above the stirring bar. Set the selection switch on the amplifier box of the probe to the 0-2000 µS/cm range.

5. Connect the Conductivity Probe to the computer interface. Prepare the computer for data collection by opening the file "26a Conductivity Eq Point" from the *Chemistry with Vernier* folder of Logger*Pro*.

6. Before adding H_2SO_4 titrant, click ▶ Collect and monitor the displayed conductivity value (in µS/cm). Once the conductivity has stabilized, click ⊛ Keep. In the edit box, type **0**, the current buret reading in mL. Press ENTER to store the first data pair for this experiment.

7. You are now ready to begin the titration. This process goes faster if one person manipulates and reads the buret while another person operates the computer and enters volumes.

    a. Add 1.0 mL of 0.02 M H_2SO_4 to the beaker. When the conductivity stabilizes, again click ⊕ Keep . In the edit box, type the current buret reading. Press ENTER. You have now saved the second data pair for the experiment.

    b. Continue adding 1.0 mL increments of H_2SO_4, each time entering the buret reading, until the conductivity has dropped below 100 μS/cm.

    c. After the conductivity has dropped below 100 μS/cm, add one 0.5 mL increment and enter the buret reading.

    d. After this, use 2-drop increments (~0.1 mL) until the minimum conductivity has been reached at the equivalence point. Enter the volume after each 2-drop addition. When you have passed the equivalence point, continue using 2-drop increments until the conductivity is greater than 50 μS/cm again.

    e. Now use 1.0 mL increments until the conductivity reaches about 2000 μS/cm, or 25 mL of H_2SO_4 solution have been added, whichever comes first.

8. When you have finished collecting data, click ■ Stop . Dispose of the beaker contents as directed by your teacher.

9. Print a copy of the table.

10. Print a copy of the graph.

METHOD 2: Measuring Volume with a Drop Counter

1. Obtain and wear goggles.

2. Connect the Conductivity Probe to CH 1 of the computer interface. Set the selection switch on the amplifier box of the probe to the 0-2000 μS/cm range. Lower the Drop Counter onto a ring stand and connect its cable to DIG/SONIC 1.

3. Measure out 25.0 mL of $Ba(OH)_2$ solution of unknown concentration using a 100 mL graduated cylinder. Transfer the solution to a clean, dry 100 mL beaker. Then add 60 mL of distilled water to the beaker. **CAUTION:** *$Ba(OH)_2$ is toxic. Handle it with care.*

4. Measure out approximately 40 mL of ~0.02 M H_2SO_4 solution into a 250 mL beaker. Record the precise H_2SO_4 concentration in your data table. **CAUTION:** *H_2SO_4 is a strong acid, and should be handled with care.*

5. Obtain the plastic 60 mL reagent reservoir. **Note:** The bottom valve will be used to open or close the reservoir, while the top valve will be used to finely adjust the flow rate. For now, close both valves by turning the handles to a horizontal position.

    Rinse it with a few mL of the 0.02 M H_2SO_4 solution. Use a utility clamp to attach the reagent reservoir to the ring stand. Add 30 mL of 0.02 M H_2SO_4 solution to the reagent reservoir.

    Drain a small amount of H_2SO_4 solution into the beaker so it fills the reservoir's tip. To do this, turn both valve handles to the vertical position for a moment, then turn them both back to horizontal.

6. Prepare the computer for data collection by opening the file "26b Conduct (Drop Count)" from the *Chemistry with Vernier* folder.

7. To calibrate drops so that a precise volume of titrant is recorded in units of milliliters:
    a. From the Experiment menu, choose Calibrate ▶ DIG 1: Drop Counter (mL).
    b. Proceed by one of these two methods:
        • If you have previously calibrated the drop size of your reagent reservoir and want to continue with the same drop size, select the Manual button, enter the number of Drops/mL, and click [OK]. Then proceed directly to Step 8.
        • If you want to perform a new calibration, select the Automatic button, and continue with Step c below.
    c. Place a 10 mL graduated cylinder directly below the slot on the Drop Counter, lining it up with the tip of the reagent reservoir.
    d. Open the bottom valve on the reagent reservoir (vertical). Keep the top valve closed (horizontal).
    e. Click the Start button.
    f. Slowly open the top valve of the reagent reservoir so that drops are released at a slow rate (~1 drop every two seconds). You should see the drops being counted on the computer screen.
    g. When the volume of H_2SO_4 solution in the graduated cylinder is between 9 and 10 mL, close the bottom valve of the reagent reservoir.
    h. Enter the precise volume of H_2SO_4 (read to the nearest 0.1 mL) in the edit box. Record the number of Drops/mL displayed on the screen for possible future use.
    i. Click [OK]. Discard the H_2SO_4 solution in the graduated cylinder as indicated by your instructor and set the graduated cylinder aside.

8. Assemble the apparatus.
    a. Place the magnetic stirrer on the base of the ring stand.
    b. Insert the Conductivity Probe through the large hole in the Drop Counter.

    c. Attach the Microstirrer to the bottom of the Conductivity Probe, as shown in the small picture. Rotate the paddle wheel of the Microstirrer and make sure that it does not touch the bottom of the Conductivity Probe.
    d. Adjust the positions of the Drop Counter and reagent reservoir so they are both lined up with the center of the magnetic stirrer.

    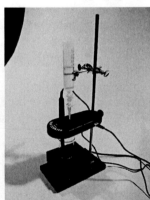

    e. Lift up the Conductivity Probe, and slide the beaker containing the $Ba(OH)_2$ solution onto the magnetic stirrer. Lower the Conductivity Probe into the beaker.
    f. Adjust the position of the Drop Counter so that the Microstirrer on the Conductivity Probe is just touching the bottom of the beaker.
    g. Adjust the reagent reservoir so its tip is just above the Drop Counter slot.

9. Turn on the magnetic stirrer so that the Microstirrer is stirring at a fast rate.

10. You are now ready to begin collecting data. Click ▶ Collect. No data will be collected until the first drop goes through the Drop Counter slot. Fully open the bottom valve—the top valve should still be adjusted so drops are released at a rate of about 1 drop every 2 seconds. When the first drop passes through the Drop Counter slot, check the data table to see that the first data pair was recorded.

11. Continue watching your graph to see when the conductivity has reached a minimum value—this will be the equivalence point of the reaction. After this minimum conductivity occurs, let the titration proceed until the conductivity reading is about the same as the initial conductivity value, then click ■ Stop. Turn the bottom valve of the reagent reservoir to a closed (horizontal) position.

12. Dispose of the beaker contents as directed by your teacher.

13. Print copies of the table.

14. Print copies of the graph.

PROCESSING THE DATA

1. From the data table and graph that you printed, determine the volume of H_2SO_4 added at the equivalence point. The graph should give you the *approximate* volume at this point. The *precise* volume of H_2SO_4 added can be determined by further examination of the data table for the minimum conductivity. Record the volume of H_2SO_4.

2. Calculate moles of H_2SO_4 added at the equivalence point. Use the molarity, M, of the H_2SO_4 and its volume, in L.

3. Calculate the moles of $Ba(OH)_2$ at the equivalence point. Use your answer in the previous step and the ratio of moles of $Ba(OH)_2$ and H_2SO_4 in the balanced equation (or use the 1:1 ratio of moles of H^+ to moles of OH^- from the equation).

4. From the moles and volume of $Ba(OH)_2$, calculate the concentration of $Ba(OH)_2$, in mol/L.

EQUIVALENCE POINT DETERMINATION: An Alternate Method

An alternate way of determining the precise equivalence point of this titration is to perform two linear regressions on the data. One of these will be on the linear region of data approaching the equivalence point, and the other will be the linear region of data following the equivalence point. The equivalence point volume corresponds to the volume at the intersection of these two lines.

1. Drag your mouse cursor across the linear region of data that *precedes* the minimum conductivity reading. Click on the Linear Fit button, ⊠.

2. Drag your mouse cursor across the linear region of data that *follows* minimum conductivity reading. Click on the Linear Fit button, ⊠.

3. Choose Interpolate from the Analyze menu. Then move the mouse cursor to the volume reading when both linear fits display the same conductivity reading. This volume reading will correspond to the equivalence point volume for the titration.

DATA TABLE

Molarity of H_2SO_4		M
Volume of H_2SO_4	mL =	L
Volume of $Ba(OH)_2$	mL =	L

Moles of H_2SO_4	
	mol
Moles of $Ba(OH)_2$	
	mol
Molarity of $Ba(OH)_2$	
	M

Using Conductivity to Find an Equivalence Point

1. The student pages with complete instructions for data-collection using LabQuest App, Logger *Pro* (computers), EasyData or DataMate (calculators), and DataPro (Palm handhelds) can be found on the CD that accompanies this book. See *Appendix A* for more information.

2. Preparation of 0.020 M H_2SO_4:

    Add 1.10 mL of concentrated H_2SO_4 per 1 L of solution. For better accuracy, you can first prepare 0.20 M H_2SO_4 by adding 11.0 mL of concentrated H_2SO_4 per 1 L of solution; then combine 100 mL of the 0.20 M H_2SO_4 with 900 mL of water to produce 1 L of 0.020 M H_2SO_4. **HAZARD ALERT:** Sulfuric acid is severely corrosive to eyes, skin and other tissue; considerable heat of dilution with water; mixing with water may cause spraying and spattering. Solutions might best be made by immersing the mixing vessel in an ice bath. **Always add the acid to water, never the reverse**; extremely hazardous in contact with finely divided materials, carbides, chlorates, nitrates and other combustible materials. Hazard Code: A—Extremely hazardous.

    To take into account small variations in the molarity of concentrated H_2SO_4, we recommend you use one of the following methods. The easiest is to assume that the solution is precisely 0.0200 M, and have your students do their experiment and calculations based on this value. This will yield a very consistent molarity value for the $Ba(OH)_2$ solution of unknown concentration (though it may deviate slightly from the actual value). A second option is to standardize the H_2SO_4 using a standard NaOH solution. This involves extra time on your part, and may not significantly improve the results of the experiment.

3. Preparation of $Ba(OH)_2$ of unknown concentration (~ 0.006 M)

    Add 2.00 g of solid $Ba(OH)_2 \cdot 8H_2O$ per 1 L of solution. **HAZARD ALERT:** Barium hydroxide is toxic by ingestion. Hazard Code: B—Hazardous. Solubility values suggest that this amount of $Ba(OH)_2$ should dissolve to produce a 0.00635 M solution. However, dissolved Ba^{2+} ions will precipitate with CO_3^{2-} and HCO_3^- ions that result from dissolved CO_2. As a result, the concentration of $Ba(OH)_2$ found in this experiment will be considerably less than 0.00635 M. The solution should be filtered on the day of the lab. This will remove any precipitates, and ensure that the resulting solution is homogeneous. If very small amounts of the precipitate appear, another option is to simply shake the solution well before dispensing. In order for students to get consistent results, they must be titrating a *homogeneous* mixture. The filtration and/or shaking are very important for consistent results.

    The hazard information reference is: Flinn Scientific, Inc., *Chemical & Biological Catalog Reference Manual*, (800) 452-1261, www.flinnsci.com. See *Appendix D* of this book.

4. You can have your students add 2–3 drops of phenolphthalein indicator to the $Ba(OH)_2$ solution at the beginning of the titration. It is interesting for them to see the indicator turn from red to clear at the same time the conductivity reaches a minimum value.

5. There are two methods described in the student procedure. Method 1 has the student deliver volumes of H_2SO_4 titrant from a buret, so buret volumes must be read (and entered) manually. Method 2 uses a Vernier Drop Counter to take volume readings. Sample data shown on the following pages are for Method 1.

6. For a more accurate equivalence-point determination, students can apply a linear fit on the linear points *before* the equivalence point and a second linear fit on linear data *after* the equivalence point. They can then determine the equivalence point by interpolating on *both* fits. Directions are in the section, Equivalence Point Determination: An Alternate Method.

SAMPLE RESULTS

Volume mL	Conductivity μS	Volume mL	Conductivity μS	Volume mL	Conductivity μS
0.0	967.9	12.0	117.0	14.5	38.6
1.0	907.3	13.0	55.3	15.0	64.7
2.0	835.1	13.5	29.2	16.0	160.9
3.0	748.4	13.6	20.8	17.0	257.1
4.0	677.3	13.7	16.6	18.0	351.1
5.0	606.2	13.8	13.5	19.0	451.5
6.0	533.0	13.9	12.4	20.0	539.3
7.0	461.9	14.0	13.5	21.0	627.1
8.0	391.9	14.1	14.5	22.0	712.8
9.0	324.0	14.2	16.6	23.0	804.8
10.0	252.9	14.3	25.0	24.0	888.4
11.0	183.9	14.4	31.3	25.0	968.9

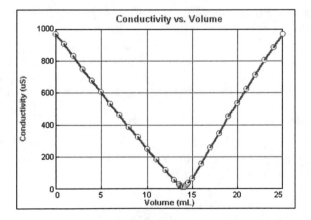

DATA AND CALCULATIONS

Molarity of H_2SO_4	0.0200 M
Volume of H_2SO_4	13.9 mL = 0.0139 L
Volume of $Ba(OH)_2$	50.0 mL = 0.0500 L

Moles of H_2SO_4 $(0.0200 \text{ mol/L})(0.0139 \text{ L}) =$ 0.000278 mol
Moles of $Ba(OH)_2$ $(0.000278 \text{ mol } H_2SO_4)(1 \text{ mol } Ba(OH)_4/1 \text{ mol } H_2SO_4) =$ 0.000278 mol
Molarity of $Ba(OH)_2$ $0.000278 \text{ mol}/0.0500 \text{ L} =$ 0.00556 mol/L

Acid Dissociation Constant, K_a

Acetic Acid, $HC_2H_3O_2$, is a weak acid that dissociates according to this equation:

$$HC_2H_3O_2(aq) \longrightarrow H^+(aq) + C_2H_3O_2^-(aq)$$

In this experiment, you will experimentally determine the dissociation constant, K_a, for acetic acid, starting with solutions of different initial concentrations.

OBJECTIVES

In this experiment, you will

- Gain experience mixing solutions of specified concentration.
- Experimentally determine the dissociation constant, K_a, of an acid.
- Investigate the effect of initial solution concentration on the equilibrium constant.

MATERIALS

computer	wash bottle
Vernier computer interface	distilled water
Logger*Pro*	100 mL volumetric flask
Vernier pH Sensor	pipets
100 mL beaker	pipet bulb
2.00 M $HC_2H_3O_2$	

PRE-LAB

1. Write the equilibrium constant expression, K_a, for the dissociation of acetic acid, $HC_2H_3O_2$. (Use Space 3 in the Data and Calculations table of this experiment.)

2. You have been assigned two different $HC_2H_3O_2$ solution concentrations by your teacher. Determine the volume, in mL, of 2.00 M $HC_2H_3O_2$ required to prepare each. (Show your calculations and answers in Space 4 of the Data and Calculations table.)

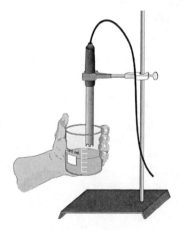

Figure 1

PROCEDURE

1. Obtain and wear safety goggles.

2. Put approximately 50 mL of distilled water into a 100 mL volumetric flask.

3. Use a pipet bulb (or pipet pump) to pipet the required volume of 2.00 M acetic acid (calculated in Pre-Lab Step 2) into the volumetric flask. **CAUTION:** *Use care when handling the acetic acid. It can cause painful burns if it comes in contact with your skin or gets into your eyes.* Fill the flask with distilled water to the 100 mL mark. To prevent overshooting the mark, use a wash bottle filled with distilled water for the last few mL. Mix thoroughly.

4. Use a utility clamp to secure a pH Sensor to a ring stand as shown in Figure 1.

5. Connect the probe to the computer interface. Prepare the computer for data collection by opening the file "27 Acid Dissociation Ka" from the *Chemistry with Vernier* folder of Logger*Pro*.

6. Determine the pH of your solution as follows:

   - Use about 40 mL of distilled water in a 100 mL beaker to rinse the pH Sensor.
   - Pour about 30 mL of your solution into a clean 100 mL beaker and use it to thoroughly rinse the sensor.
   - Repeat the previous step by rinsing with a second 30 mL portion of your solution.
   - Use the remaining 40 mL portion to determine pH. Swirl the solution vigorously. (Note: Readings may drift without proper swirling!) Record the measured pH reading in your data table.
   - When done, place the pH Sensor in distilled water.

7. Repeat the procedure for your second assigned solution.

PROCESSING THE DATA

1. Use a scientific calculator to determine the $[H^+]_{eq}$ from the pH values for each solution.

2. Use the obtained value for $[H^+]_{eq}$ and the equation:

$$HC_2H_3O_2(aq) \longrightarrow H^+(aq) + C_2H_3O_2^-(aq)$$

   to determine $[C_2H_3O_2^-]_{eq}$ and $[HC_2H_3O_2]_{eq}$.

3. Substitute these calculated concentrations into the K_a expression you wrote in Step 1 of the Pre-Lab.

4. Compare your results with those of other students. What effect does initial $HC_2H_3O_2$ concentration seem to have on K_a?

OBSERVATIONS

DATA TABLE

1. Assigned concentration	M	M
2. Measured pH		
3. K_a expression		
4. Volume of 2 M acetic acid	mL	mL
5. $[H^+]_{eq}$	M	M
6. $[C_2H_3O_2^-]_{eq}$	M	M
7. $[HC_2H_3O_2]_{eq}$	M	M
8. K_a calculation		

Acid Dissociation Constant, K$_a$

1. The student pages with complete instructions for data-collection using LabQuest App, Logger *Pro* (computers), EasyData or DataMate (calculators), and DataPro (Palm handhelds) can be found on the CD that accompanies this book. See *Appendix A* for more information.

2. Prepare adequate 2.00 M HC$_2$H$_3$O$_2$ (113.8 mL concentrated acetic acid, HC$_2$H$_3$O$_2$, per 1 L solution). A class of 24 students working in teams of two would require about 500 mL. **HAZARD ALERT:** Corrosive to skin and tissue; moderate fire risk (flash point: 39°C); moderately toxic by ingestion and inhalation. Hazard Code: A—Extremely hazardous.

   The hazard information reference is: Flinn Scientific, Inc., *Chemical & Biological Catalog Reference Manual*, (800) 452-1261, www.flinnsci.com. See *Appendix D* of this book, *Chemistry with Vernier*, for more information.

3. Assign each lab team two acetic acid solution concentrations the day before the experiment is to be done. Concentrations of 0.200 M, 0.300 M, 0.400 M and 0.500 M will require 10.0, 15.0, 20.0 and 25.0 mL, respectively, of 2.00 M HC$_2$H$_3$O$_2$.

4. If you have an insufficient quantity of 100 mL volumetric flasks, 100 mL graduated cylinders can be substituted.

5. Pipets of various sizes will be needed. Be sure to provide appropriate pipet bulbs or pipet pumps. Graduated 10 mL disposable pipets work well for most solutions.

6. You may want to use another acid or acids. The use of pipets can be eliminated with the choice of a water soluble, solid acid.

7. It is important to have a well-calibrated pH Sensor for this experiment. Using the Logger *Pro* experiment file for Experiment 27, check a pH Sensor in pH-7 buffer to see if it gives a reading close to pH 7. If it does, then use the calibration that is stored with the experiment file for the experiment. If not, use buffers of pH 4.0 and 7.0 to calibrate the pH Sensor, following these steps:

   First Calibration Point

   a. Set up the data-collection software to calibrate the pH Sensor.
   b. For the first calibration point, rinse the pH Sensor with distilled water, then place it into a buffer of pH 4.0.
   c. Type **4** in the edit box as the pH value.
   d. Swirl the sensor, wait until the voltage stabilizes, and Keep the point.

   Second Calibration Point

   e. Rinse the pH Sensor with distilled water, and place it into a buffer of pH 7.0.
   f. Type **7** in the edit box as the pH value for the second calibration point.
   g. Swirl the sensor and wait until the voltage stabilizes. Keep the point, and then click select OK or [**Done**] depending on the software. This completes the calibration.
   h. You are now ready to collect data, using the calibration. Or, to use the calibration values for a specific pH Sensor at a later time, you can simply resave this experiment file—the new calibration is saved along with the experiment file itself.

8. The computer procedure has students record pH values from the live readouts(without starting data collection). Another possibility is to have students use the Selected Events mode for each of the trials. In Logger*Pro,* the file for this experiment is already set up for this option. Simply have your students click ▶ Collect , then click ⊛ Keep when the pH reading is stable. This saves the pH reading along with its trial number.

SAMPLE RESULTS

1. Assigned concentration	0.400 M
2. Measured pH	2.57
3. K_a expression	$$K_a = \frac{[H^+]\,[C_2H_3O_2^-]}{[HC_2H_3O_2]}$$
4. Volume of 2 M acetic acid	$(0.100\ L) \times (0.400\ mol/L) = 0.0400\ mol$ $(0.0400\ mol) \times (1\ L/2.00\ mol) = 0.0200\ L$ 20.0 mL
5. $[H^+]_{eq}$	$2.57 = -\log [H^+]$ $[H^+] = 2.69 \times 10^{-3}\ M$ $2.69 \times 10^{-3}\ M$
6. $[C_2H_3O_2^-]_{eq}$	$[C_2H_3O_2^-] = [H^+] = 2.69 \times 10^{-3}\ M$ $2.69 \times 10^{-3}\ M$
7. $[HC_2H_3O_2]_{eq}$	$0.400 - 0.003 = 0.397\ M$ 0.397 M
8. K_a calculation	$\dfrac{(2.69 \times 10^{-3})^2}{(0.397)} = 1.82 \times 10^{-5}$ 1.82×10^{-5}

Establishing a Table of Reduction Potentials: Micro-Voltaic Cells

The main objective of this experiment is to establish the reduction potentials of five unknown metals relative to an arbitrarily chosen metal. This will be done by measuring the voltage, or potential difference, between various pairs of half-cells.

A voltaic cell utilizes a spontaneous oxidation-reduction reaction to produce electrical energy. Half-cells are normally produced by placing a piece of metal into a solution containing a cation of the metal (e.g., Cu metal in a solution of $CuSO_4$ or Cu^{2+}). In this micro-version of a voltaic cell, the half cell will be a small piece of metal placed into 3 drops of solution on a piece of filter paper. The solution contains the cation of the solid metal. Figure 1 shows the arrangement of half-cells on the piece of filter paper. The two half-reactions are normally separated by a porous barrier or a salt bridge. Here, the salt bridge will be several drops of aqueous $NaNO_3$ placed on the filter paper between the two half cells. Using the computer as a voltmeter, the (+) lead makes contact with one metal and the (−) lead with another. If a positive voltage is recorded on the screen, you have connected the cell correctly. The metal attached to the (+) lead is the cathode (reduction) and thus has a higher, more positive, reduction potential. The metal attached to the (−) lead is the anode (oxidation) and has the lower, more negative, reduction potential. If you get a negative voltage reading, then you must reverse the leads.

By comparing the voltage values obtained for several pairs of half-cells, and by recording which metal made contact with the (+) and (−) leads, you can establish the reduction potential sequence for the five metals in this lab.

OBJECTIVES

In this experiment, you will establish the reduction potentials of five unknown metals relative to an arbitrarily chosen metal.

MATERIALS

computer	one piece of filter paper, 11.0 cm diameter
Vernier computer interface	1×1 cm metals M_1, M_2, M_3, M_4 and M_5
LoggerPro	1 M $NaNO_3$
Vernier Voltage Probe	1 M solutions of M_1^{2+}, M_2^{2+}, ..., and M_5^{2+}
one glass plate, 15 × 15 cm, or one Petri dish, 11.5 cm diameter	sand paper
	forceps

PROCEDURE

1. Obtain and wear goggles.

2. Connect the Voltage probe to the computer interface. Prepare the computer for data collection by opening the file "28 Micro-voltaic Cells" from the *Chemistry with Vernier* folder of LoggerPro. **Note:** When the voltage measurement leads are not in contact with a cell (or each other), a meaningless default voltage may be displayed. If you touch the two leads together, the voltage with drop to about 0.00 V.

3. Obtain a piece of filter paper and draw five small circles with connecting lines, as shown in Figure 1. Using a pair of scissors, cut wedges between the circles as shown. Label the circles M_1, M_2, M_3, M_4, and M_5. Place the filter paper on top of the glass plate.

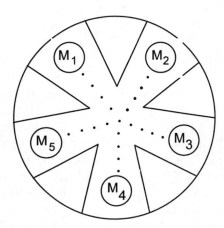

Figure 1

4. Obtain 5 pieces of metal, M_1, M_2, M_3, M_4, and M_5. Sand both surfaces of each piece of metal. Place each metal near the circle with the same number.

5. Place 3 drops of each solution on its circle (M_1^{2+} on M_1, etc.). Then place the piece of metal on the wet spot with its respective cation. The top side of the metal should be kept dry. Then add several drops of 1 M $NaNO_3$ to the line drawn between each circle and the center of the filter paper. Be sure there is a continuous trail of $NaNO_3$ between each circle and the center. You may have to periodically dampen the filter paper with $NaNO_3$ during the experiment. **CAUTION:** *Handle these solutions with care. Some are poisonous and some cause hard-to-remove stains. If a spill occurs, ask your teacher how to clean up safely.*

6. Use metal M_1 (the one that is obviously copper) as the reference metal. Determine the potential of four cells by connecting M_1 to M_2, M_1 to M_3, M_1 to M_4, and M_1 to M_5. This is done by bringing the (+) lead in contact with one metal and the (−) lead in contact with the other. If the voltage displayed in the meter is negative then reverse the leads.

   With a positive voltage displayed, wait about 5 seconds to take a voltage reading, and record the value in Data Table 1. Also record which metal is the (+) terminal and which is (−). Use the same procedure and measure the potential of the other three cells, continuing to use M_1 as the reference electrode.

7. Go to Step 1 of Processing the Data. Use the method described in Step 1 to rank the five metals from the lowest (−) reduction potential to the highest (+) reduction potential. Then *predict* the potentials for the remaining six cell combinations.

8. Using the computer and Voltage Probe, measure the potential of the six remaining half-cell combinations. If the $NaNO_3$ salt bridge solution has dried, you may have to re-moisten it. Record each measured potential in Data Table 3.

9. When you have finished, use forceps to remove each of the pieces of metal from the filter paper. Rinse each piece of metal with tap water. Dry it and return it to the correct container. Remove the filter paper from the glass plate using the forceps, and discard it as directed by your teacher. Rinse the glass plate with tap water, making sure that your hands do not come in contact with wet spots on the glass.

PROCESSING THE DATA

1. After finishing Step 6 in the procedure, arrange the five metals (including M_1) in Data Table 2 from the lowest reduction potential at the top (most negative) to the highest reduction potential at the bottom (most positive). Metal M_1, the standard reference, will be given an arbitrary value of 0.00 V. If the other metal was correctly connected to the *negative* terminal, it will be placed *above* M_1 in the chart (with a negative $E°$ value). If it was connected to the positive terminal, it will be placed below M_1 in the chart (with a positive $E°$ value). The numerical value of the potential relative to M_1 will simply be the value that you measured on the computer. Record your results in Data Table 2.

   Then calculate the predicted potential of each of the remaining cell combinations shown in Data Table 3, using the reduction potentials you just determined (in Data Table 2). Record the predicted cell potentials in Data Table 3. Return to Step 8 in the procedure and finish the experiment.

2. Calculate the % error for each of the potentials you measured in Step 8 of the procedure. Do this by comparing the measured cell potentials with the predicted cell potentials in Data Table 3.

3. (Optional) You can determine the identity of metals M_2 through M_5 using a reduction potential chart in your textbook. Remember that hydrogen, H_2, has a reduction potential of 0.00 V on this chart. Locate copper, M_1, on the chart, and then determine the likely identity of each of the other metals using your experimental reduction potential sequence in Data Table 2. Note: One of the metals has a 1^+ oxidation state; the remainder of the metals have 2^+ oxidation states.

DATA TABLE 1

Voltaic Cell (metals used)	Measured Potential (V)	Metal Number of (+) Lead	Metal Number of (−) Lead
M_1 / M_2			
M_1 / M_3			
M_1 / M_4			
M_1 / M_5			

DATA TABLE 2

Metal (M_X)	Lowest (−) Reduction Potential, $E°$ (V)
	Highest (+) Reduction Potential, $E°$ (V)

DATA TABLE 3

	Predicted Potential (V)	Measured Potential (V)	Percent Error (%)
M_2 / M_3			
M_2 / M_4			
M_2 / M_5			
M_3 / M_4			
M_3 / M_5			
M_4 / M_5			

Establishing a Table of Reduction Potentials: Micro-Voltaic Cells

1. The student pages with complete instructions for data-collection using LabQuest App, Logger *Pro* (computers), EasyData or DataMate (calculators), and DataPro (Palm handhelds) can be found on the CD that accompanies this book. See *Appendix A* for more information.

2. When the voltage measurement leads are not in contact with a cell (or each other), a meaningless default voltage may be displayed. If you touch the two leads together, the voltage with drop to about 0.00 V.

3. The micro-scale dimensions of this lab provide tremendous savings in quantities of reagents used in preparing 1.0 M solutions. Each student uses approximately 0.15 mL of each solution in preparing the voltaic cells; thus the total amount of solution used for a class is less than 5 mL. A 30 mL dropper bottle of solution can be expected to last for several years. The micro-scale dimensions also allow you to use silver metal and 1.0 M $AgNO_3$ solution. The cost of such a macro-scale cell would normally be prohibitive.

4. Solutions are prepared using the quantities shown below. Use distilled water for all solutions. The amounts shown will be enough to fill three sets of 30 mL dropper bottles.

    1.0 M $CuSO_4$ (M_1^{2+}) (24.96 g solid $CuSO_4 \cdot 5H_2O$ per 100 mL) **HAZARD ALERT:** Skin and respiratory irritant; toxic by ingestion and inhalation. Code: C—Somewhat hazardous.

    1.0 M $ZnSO_4$ (M_2^{2+}) (28.76 g solid $ZnSO_4 \cdot 7H_2O$ per 100 mL) **HAZARD ALERT:** Skin and mucus membrane irritant: moderately toxic. Hazard Code: D—Relatively non-hazardous.

    1.0 M $Pb(NO_3)_2$ (M_3^{2+}) (33.10 g solid $Pb(NO_3)_2$ per 100 mL) **HAZARD ALERT:** Toxic by inhalation and ingestion; strong oxidant; dangerous fire risk in contact with organic material. *Possible carcinogen.* Hazard Code: B—Hazardous.

    1.0 M $AgNO_3$ (M_4^+) (16.99 g solid $AgNO_3$ per 100 mL) **HAZARD ALERT:** Corrosive solid; causes burns; avoid contact with eyes and skin; toxic. Hazard Code: B—Hazardous.

    1.0 M $FeSO_4$ (M_5^{2+}) (27.80 g solid $FeSO_4 \cdot 7H_2O$ per 100 mL) **HAZARD ALERT:** Moderately toxic by ingestion. Hazard Code: C—Somewhat hazardous.

    1.0 M $NaNO_3$ (8.50 g solid $NaNO_3$ per 100 mL) **HAZARD ALERT:** Strong oxidizer; avoid friction or shock—explosions have occurred; moderately toxic by ingestion (B—Hazardous).

    The hazard information reference is: Flinn Scientific, Inc., *Chemical & Biological Catalog Reference Manual,* (800) 452-1261, www.flinnsci.com. See *Appendix D* of this book, *Chemistry with Vernier*, for more information.

5. All of these solutions will store well except the 1.0 M $FeSO_4$. It is recommended that this solution be freshly prepared each year. The 1.0 M $AgNO_3$ solution should be stored in an opaque or brown-glass bottle. It should also be kept in a dark cabinet when not in use.

6. The pieces of metal should be cut in sizes approximately 1 cm × 1 cm. To prevent students from returning the metals to the wrong containers, we recommend that each metal be cut in its own distinct shape (square, triangle, trapezoid). The metals can be keyed as follows:

   M_1 = Cu; M_2 = Zn; M_3 = Pb; M_4 = Ag; M_5 = Fe. As an alternative to a glass plate, a large-diameter Petri dish also works well.

7. The computer procedure has students record voltage values from the live readouts, without starting data collection. Another possibility is to have students use the Selected Events mode for each of the 10 trials. In Logger*Pro,* the file for this experiment is already set up for this option. Simply have your students click ▶ Collect , click ⊛ Keep when the voltage reading is stable. This saves the voltage reading along with its trial number.

SAMPLE RESULTS
DATA TABLE 1

Voltaic Cell (metals used)	Measured potential (V)	Metal number of (+) Lead	Metal number of (-) Lead
M_1 / M_2	1.08 V	M_1	M_2
M_1 / M_3	0.47 V	M_1	M_3
M_1 / M_4	0.45 V	M_4	M_1
M_1 / M_5	0.64 V	M_1	M_5

DATA TABLE 2

Metal (M_X)	Lowest (-) reduction potential, E° (V)
M_2 (Zn/Zn^{2+})	–1.08 V
M_5 (Fe/Fe^{2+})	–0.64 V
M_3 (Pb/Pb^{2+})	–0.47 V
M_1 (Cu/Cu^{2+})	0.00 V
M_4 (Ag/Ag^+)	+0.45 V
	Highest (+) reduction potential, E° (V)

DATA TABLE 3

	Predicted potential (V)	Measured potential (V)	Percent error (%)
M_2 / M_3	–0.47– (–1.08) = 0.61 V	0.60 V	1.6 %
M_2 / M_4	0.45– (–1.08) = 1.53 V	1.49 V	2.6 %
M_2 / M_5	–0.64– (–1.08) = 0.44 V	0.46 V	4.5 %
M_3 / M_4	0.45– (–0.47) = 0.92 V	0.90 V	2.2 %
M_3 / M_5	–0.47– (–0.64) = 0.17 V	0.16 V	5.9 %
M_4 / M_5	0.45– (–0.64) = 1.09 V	1.07 V	1.8 %

RESULTS FROM OPTIONAL SECTION

The results in Table 2 are all based on the use of Cu as the reference electrode with a standard reduction potential of 0.00 V. If hydrogen, H_2, is used instead, as in the Table of Standard Reduction Potentials, 0.34 V must be added to each of the values obtained in Table 2. This places Cu in its usual position of +0.34 V. Values obtained in this lab are listed in the second column below. Values from the Table of Standard Reduction Potentials are shown in the third column. With the possible exception of iron, all of the metals can be easily identified from these values. Iron will likely be mistaken for Ni (E° for Ni = –0.28 V).

Metal	Experimental E°	Standard E°
M_2 (Zn)	–0.74 V	–0.76 V
M_5 (Fe)	–0.30 V	–0.44 V
M_3 (Pb)	–0.13 V	–0.13 V
M_1 (Cu)	+0.34 V	+0.34 V
M_4 (Ag)	+0.79 V	+0.80 V

Lead Storage Batteries

Two or more wet or dry cells connected in series make a battery. A car battery is generally a lead storage battery containing lead and lead oxide plates in sulfuric acid solution. In this experiment, you will construct a lead storage cell and use a direct-current power supply to charge it as shown in Figure 1. You will use the computer and a Voltage Probe to measure the cell's voltage (see Figure 2), and then use the cell to power an electric motor.

OBJECTIVES

In this experiment, you will

- Construct a lead storage cell.
- Use a Voltage Probe to measure a cell's voltage.
- Use the cell to power an electric motor.

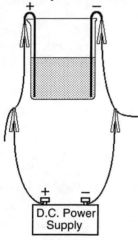

Figure 1

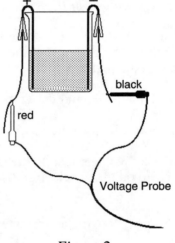

Figure 2

MATERIALS

computer	250 mL beaker
Vernier computer interface	2 alligator clips
Logger *Pro*	sulfuric acid, H_2SO_4
Vernier Voltage Probe	clock (with second hand)
direct-current power supply	small electric motor
2 lead strips (2 cm × 12 cm)	2 test leads
apron	

PROCEDURE

1. Obtain and wear goggles and an apron. **CAUTION:** The battery acid, H_2SO_4, used in this experiment can damage eyesight and make holes in clothing!

2. Obtain two lead strips. If the strips have been used before, get one labeled (+) and one labeled (–). If the strips are not marked, label one (+) and the other (–). Bend the strips and place them in a 250 mL beaker as shown in Figure 1. Attach an alligator clip to each lead strip.

3. Add 125 mL of sulfuric acid, H_2SO_4, to the beaker. Handle this strong acid with care!

4. Connect the Voltage probe to the computer interface. Prepare the computer for data collection by opening the file "29 Lead Batteries" from the *Chemistry with Vernier* folder of Logger*Pro*.

5. Charge the cell using the direct-current power supply:

    a. Attach the (–) lead from the power supply to the alligator clip on the (–) Pb electrode as shown in Figure 1.
    b. Attach the (+) lead of the power supply to the alligator clip on the (+) Pb electrode to begin the charging process.
    c. Time the charging process and disconnect the power supply leads after 4 minutes. Record observations during the charging process. **CAUTION:** *Make sure the lead strips do not touch each other while connected to the power supply.*

6. Attach the *red clip* of the Voltage Probe to the alligator clip on the (+) electrode (the black clip should still be attached to the (–) electrode via an alligator clip). Read the voltage value displayed in the meter. Record the reading after it stabilizes.

7. Disconnect the black and red voltage leads from the cell. Use two wire leads to connect the cell to a small electric motor. Use a clock to measure the number of seconds the charged cell runs the motor. Record the results. The cell is said to be *discharging* during this process.

8. Attach the red voltage lead to the alligator clip on the (+) electrode and its black clip to the alligator clip on the (–) electrode. Measure the voltage of the *discharged* cell. Record this value.

9. Repeat Steps 5–8 using a 2-minute charging time.

10. Observe the two lead electrodes and record your observations.

11. Return the H_2SO_4 solution to the "Used H_2SO_4" container supplied by your instructor. Wash and dry the beaker and the lead strips.

PROCESSING THE DATA

1. From the voltage values for the 1st and 2nd charging, calculate the average voltage of your cell when charged.

2. Cars generally have 12 volt batteries. How many lead storage cells, similar to the one you built, does a car battery contain? Explain.

3. Using a Table of Standard Reduction Potentials, write the equation occurring at the anode and cathode when the battery was discharging and behaving as a voltaic (electrochemical) cell. Write the standard potential value, $E°$, in the blank following the equation. In the third blank, write the net equation for the reaction by combining the two half-reactions. Find the $E°_{total}$ (or $E°_{cell}$) by adding the $E°$ values for the two half-reactions.

4. Find the percent error for the cell potential by comparing your experimental voltage value in Step 1 of Processing the Data with the accepted $E°_{total}$ value in Step 3.

5. What was the gas you saw being produced at the (−) electrode, during charging? What was the gas being produced at the (+) electrode? Account for the danger of an explosion after car battery charging.

6. Explain why "run-down" car batteries sometimes freeze up and break open in extremely cold weather. **Hint:** Examine the equation for the net reaction in the Data and Calculations table below.

7. When you charged and discharged the battery in this experiment, which process was electrolytic? Which was electrochemical (voltaic)? Explain.

DATA AND CALCULATIONS

	1st Charging	2nd Charging
Voltage after charging	_____ V	_____ V
Time motor ran after charging	_____ s	_____ s
Voltage after discharge	_____ V	_____ V

Average potential of charged cell
_____ V

Equation	$E°$
Anode (−) _____	_____ V
Cathode (+) _____	_____ V
Net Reaction _____	_____ V

Percent error
_____ %

OBSERVATIONS

TEACHER INFORMATION

Lead Storage Batteries

1. The student pages with complete instructions for data-collection using LabQuest App, Logger *Pro* (computers), EasyData or DataMate (calculators), and DataPro (Palm handhelds) can be found on the CD that accompanies this book. See *Appendix A* for more information.

2. Go through the instructions with your students very thoroughly. If you are concerned about student use of dilute sulfuric acid, consider doing this experiment as a demonstration.

3. Attaching the cable clips of the Voltage Probe to alligator clips, with or without wires attached, reduces corrosion of the Voltage Probe clips.

4. Use 1.0 M sulfuric acid (56 mL concentrated H_2SO_4 per 1 L). **HAZARD ALERT:** Severely corrosive to eyes, skin and other tissue; considerable heat of dilution with water; mixing with water may cause spraying and spattering. Solutions might best be made by immersing the mixing vessel in an ice bath. **Always add the acid to water, never the reverse**; extremely hazardous in contact with finely divided materials, carbides, chlorates, nitrates and other combustible materials. Hazard Code: A—Extremely hazardous.

   Place a large beaker labeled "Used H_2SO_4" out for return of the sulfuric acid after use.

5. Lead strips that are ~2 cm × 12 cm work well. If you label them with a permanent marker as either (+) or (–), then it will not be necessary to sand off the thin brown layer of PbO_2 that accumulates on the (+) lead strip prior to the next use. **HAZARD ALERT:** Lead as a powder is extremely toxic; the substance can be absorbed by inhalation and ingestion. *Possible carcinogen* as a fume or dust. Take all precautions when working with lead powder. Hazard Code (sheet): D—Relatively non-hazardous. Hazard Code (powder): A—Extremely hazardous.

   The hazard information reference is: Flinn Scientific, Inc., *Chemical & Biological Catalog Reference Manual* (800) 452-1261, www.flinnsci.com. See *Appendix D* of this book, *Chemistry with Vernier*, for more information.

6. Small motors can be obtained at Radio Shack stores. For increased effect, place propellers on the small electric motors used. Miniature light bulbs can be substituted for the electric motors.

7. The computer procedure has students record voltage values from the live readouts, without starting data collection. Another possibility is to have students use the Selected Events mode for each measurement. In Logger *Pro,* the file for this experiment is already set up for this option. Simply have your students click [▶ Collect], click [⊛ Keep] when the voltage reading is stable. This saves the voltage reading along with its trial number.

ANSWERS TO QUESTIONS

2. Car batteries (12 V) have six 2-V cells in series (6×2 V = 12 V).

5. Hydrogen gas is produced at the (–) anode by the reaction:

$$2\,H^+(aq) + 2\,e^- \longrightarrow H_2(g)$$

Oxygen gas is produced at the (+) cathode by the reaction:

$$2\,H_2O(l) \longrightarrow 4\,H^+(aq) + O_2(g) + 4\,e^-$$

6. In run-down batteries, the net reaction has been allowed to proceed toward the products. The products, $PbSO_4$ and H_2O, have far fewer dissolved ions than the reactants. Thus, the freezing point of the solution is not lowered as much as it was with H^+ and $SO_4{}^{2-}$ dissolved ions.

7. Charging is electrolytic. An external power supply is used to cause a reaction to proceed in a non-spontaneous direction. Discharging is electrochemical. The reaction proceeds spontaneously, and has a positive E°_{cell} (+2.05 V).

SAMPLE RESULTS

	1st Charging	2nd Charging
Voltage after charging	2.03 V	2.02 V
Time motor ran after charging	45 s	45 s
Voltage after discharge	1.92 V	1.92 V

Average potential of charged cell
$$\frac{2.03 + 2.02}{2} =$$
2.03 V

Equation		E°
Anode (–):	$Pb(s) + SO_4{}^{2-}(aq) \longrightarrow PbSO_4(s) + 2\,e^-$	0.36 V
Cathode (+):	$PbO_2(s) + SO_4{}^{2-}(aq) + 4\,H^+(aq) + 2\,e^- \longrightarrow PbSO_4(s) + 2\,H_2O(s)$	1.69 V
Net reaction:	$Pb(s) + PbO_2(s) + 2\,SO_4{}^{2-}(aq) + 4\,H^+(aq) \longrightarrow 2\,PbSO_4(s) + 2\,H_2O(l)$	2.05 V

Percent error
$$\frac{|2.05 - 2.03|}{|2.05|} \times 100 =$$
1.0 %

Rate Law Determination of the Crystal Violet Reaction

In this experiment, you will observe the reaction between crystal violet and sodium hydroxide. One objective is to study the relationship between concentration of crystal violet and the time elapsed during the reaction. The equation for the reaction is shown here.

A simplified (and less intimidating!) version of the equation is:

$$CV^+ \text{ (aq)} + OH^- \text{ (aq)} \rightarrow CVOH \text{ (aq)}$$

(crystal violet)   (hydroxide)

The rate law for this reaction is in the form: rate = $k[CV^+]^m[OH^-]^n$, where k is the rate constant for the reaction, m is the order with respect to crystal violet (CV^+), and n is the order with respect to the hydroxide ion. Because the hydroxide ion concentration is more than 1000 times as large as the concentration of crystal violet, $[OH^-]$ will not change appreciably during this experiment. Thus, you will find the order with respect to crystal violet (m), but not the order with respect to hydroxide (n).

As the reaction proceeds, a violet-colored reactant will be slowly changing to a colorless product. You will measure the color change with a Vernier Colorimeter or a Vernier Spectrometer. The crystal violet solution used in this experiment has a violet color, of course, thus the Colorimeter users will be instructed to use the 565 nm (green) LED. Spectrometer users will determine an appropriate wavelength based on the absorbance spectrum of the solution. We will assume that absorbance is proportional to the concentration of crystal violet (Beer's law). Absorbance will be used in place of concentration in plotting the following three graphs:

- Absorbance *vs.* time: A linear plot indicates a *zero order* reaction (k = –slope).
- ln Absorbance *vs.* time: A linear plot indicates a *first order* reaction (k = –slope).
- 1/Absorbance *vs.* time: A linear plot indicates a *second order* reaction (k = slope).

Once the order with respect to crystal violet has been determined, you will also be finding the rate constant, k, and the half-life for this reaction.

OBJECTIVES

In this experiment, you will

- Observe the reaction between crystal violet and sodium hydroxide.
- Monitor the absorbance of the crystal violet solution with time.
- Graph Absorbance *vs.* time, ln Absorbance *vs.* time, and 1/Absorbance *vs.* time.
- Determine the order of the reaction.
- Determine the rate constant, *k*, and the half-life for this reaction.

MATERIALS

computer	0.10 M sodium hydroxide, NaOH, solution
Vernier computer interface*	2.5×10^{-5} M crystal violet solution
Logger *Pro*	ice
Colorimeter or Spectrometer	two 10 mL graduated cylinders
Temperature Probe or thermometer	two 100 mL beakers
5 plastic cuvettes	50 mL beaker
1 liter beaker	watch with a second hand

*No interface is required if using a Spectrometer

PROCEDURE

Both Colorimeter and Spectrometer Users

1. Obtain and wear goggles.

2. Use a 10 mL graduated cylinder to obtain 10.0 mL of 0.10 M NaOH solution. **CAUTION:** *Sodium hydroxide solution is caustic. Avoid spilling it on your skin or clothing.* Use another 10 mL graduated cylinder to obtain 10.0 mL of 2.5×10^{-5} M crystal violet solution. **CAUTION:** *Crystal violet is a biological stain. Avoid spilling it on your skin or clothing.*

3. Prepare a *blank* by filling a cuvette 3/4 full with distilled water. To correctly use cuvettes, remember:

    - Wipe the outside of each cuvette with a lint-free tissue.
    - Handle cuvettes only by the top edge of the ribbed sides.
    - Dislodge any bubbles by gently tapping the cuvette on a hard surface.
    - Always position the cuvette so the light passes through the clear sides.

Spectrometer Users Only (Colorimeter users proceed to the Colorimeter section)

4. Use a USB cable to connect the Spectrometer to your computer. Choose New from the File menu.

5. To calibrate the Spectrometer, place the blank cuvette into the cuvette slot of the Spectrometer, choose Calibrate ▶ Spectrometer from the Experiment menu. The calibration dialog box will display the message: "Waiting 90 seconds for lamp to warm up." After 90 seconds, the message will change to "Warmup complete." Click ☐ OK ☐.

6. Determine the optimum wavelength for examining the crystal violet solution and set up the mode of data collection.

    a. Empty the blank cuvette and rinse it twice with small amounts of 2.5×10^{-5} M crystal violet solution. Fill the cuvette about 3/4 full with the crystal violet solution and place it in the spectrometer.

    b. Click ▶Collect. A full spectrum graph of the solution will be displayed. Note that one area of the graph contains a peak absorbance. Click ■ Stop to complete the analysis.

    c. To save your graph of absorbance *vs.* wavelength, select Store Latest Run from the Experiment menu.

    d. To set up the data collection mode and select a wavelength for analysis, click on the Configure Spectrometer Data Collection icon, 🔲, on the toolbar.

    e. Click Abs *vs.* Time (under the Set Collection Mode). The wavelength of maximum absorbance (λ max) will be selected. Click OK. Remove the cuvette from the spectrometer and dispose of the crystal violet solution as directed. Save the cuvette for Step 7.

    f. Proceed to Step 7.

Colorimeter Users Only

4. Connect the Colorimeter to the computer interface. Prepare the computer for data collection by opening the file "30b Rate Crystal Violet" from the *Advanced Chemistry with Vernier* folder of Logger*Pro*.

5. Open the Colorimeter lid, insert the blank, and close the lid.

6. To calibrate the Colorimeter, press the < or > button on the Colorimeter to select the wavelength of 565 nm (Green). Press the CAL button until the red LED begins to flash and then release the CAL button. When the LED stops flashing, the calibration is complete. Remove the cuvette from the Colorimeter and save it for Step 7.

Both Colorimeter and Spectrometer Users

7. *Do this quickly!* To initiate the reaction, simultaneously pour the 10 mL portions of crystal violet and sodium hydroxide solutions into a 250 mL beaker and stir the reaction mixture with a stirring rod. Empty the water from the cuvette. Rinse the cuvette twice with ~1 mL amounts of the reaction mixture, fill it 3/4 full, and place it in the device (Colorimeter or Spectrometer). Close the lid on the Colorimeter. Click ▶Collect.

8. Absorbance data will be collected for three minutes. Discard the beaker and cuvette contents as directed by your instructor.

9. Analyze the data graphically to decide if the reaction is zero, first, or second order with respect to crystal violet.

    • Zero Order: If the current graph of absorbance *vs.* time is linear, the reaction is *zero order*.

    • First Order: To see if the reaction is first order, it is necessary to plot a graph of the natural logarithm (ln) of absorbance *vs.* time. If this plot is linear, the reaction is *first order*.

    • Second Order: To see if the reaction is second order, plot a graph of the reciprocal of absorbance *vs.* time. If this plot is linear, the reaction is *second order*.

10. Follow these directions to create a calculated column, ln Absorbance, and then plot a graph of ln Absorbance *vs.* time:

    a. Choose New Calculated Column from the Data menu.

    b. Enter "ln Absorbance" as the Name, and leave the unit blank.

    c. Enter the correct formula for the column into the Equation edit box by choosing "ln" from the Function list, and selecting "Absorbance" from the Variables list. Click ⎡ **Done** ⎤.

    d. Click on the y-axis label. Choose ln Absorbance. A graph of ln absorbance *vs.* time should now be displayed. Change the scale of the graph, if necessary.

    e. Click the Linear Regression button, ⌧. Write down the slope value in your data table as the rate constant, k.

    f. Close the Linear Regression box by clicking the X in the corner of the box.

11. Follow these directions to create a calculated column, 1/Absorbance, and then plot a graph of 1/Absorbance *vs.* time:

    a. Choose New Calculated Column from the Data menu.

    b. Enter "1/Absorbance" as the Name, "1/Abs" as the Short Name, and leave the unit blank.

    c. Enter the correct formula for the column into the Equation edit box. To do this, type in "1" and "/". Then select "Absorbance" from the Variables list. In the Equation edit box, you should now see displayed: 1/"Absorbance". Click ⎡ **Done** ⎤.

    d. Click on the y-axis label. Choose 1/Absorbance and uncheck any other boxes. A graph of 1/Absorbance *vs.* time should now be displayed. To see if the relationship is linear, click the Linear Fit button, ⌧.

12. Print a copy of the graph in Steps 9-11 that was linear (Absorbance, ln Absorbance, or 1/Absorbance *vs.* time).

    a. Click the vertical-axis label of the graph.

    b. Of "Absorbance", "ln Absorbance", or "1/Absorbance", choose only the data that gave a linear plot. Click ⎡ **OK** ⎤.

    c. Print a copy of the graph. Enter your name(s) and the number of copies of the graph you want printed. Note: Be sure the linear regression curve is displayed on the graph, as well as the regression statistics box.

13. Print a copy of the table. Enter your name(s) and the number of copies of the table.

14. Optional: Print a copy of the two non-linear graphs.

PROCESSING THE DATA

1. Was the reaction zero, first, or second order, with respect to the concentration of crystal violet? Explain.

2. Calculate the rate constant, k, using the *slope* of the linear regression line for your linear curve ($k = -$slope for zero and first order and $k =$ slope for second order). Be sure to include correct units for the rate constant. Note: This constant is sometimes referred to as the *pseudo rate constant*, because it does not take into account the effect of the other reactant, OH^-.

3. Write the correct rate law expression for the reaction, in terms of crystal violet (omit OH^-).

4. Using the printed data table, estimate the half-life of the reaction; select two points, one with an absorbance value that is about half of the other absorbance value. The *time* it takes the absorbance (or concentration) to be halved is known the *half-life* for the reaction. (As an alternative, you may choose to calculate the half-life from the rate constant, k, using the appropriate concentration-time formula.)

Rate Law Determination of the Crystal Violet Reaction

1. The student pages with complete instructions for data-collection using LabQuest App, Logger *Pro* (computers), EasyData or DataMate (calculators), and DataPro (Palm handhelds) can be found on the CD that accompanies this book. See *Appendix A* for more information.

2. Preparation of solutions:

    0.10 M NaOH: 4.00 g of solid NaOH per 1 L solution. **HAZARD ALERT:** Corrosive solid; skin burns are possible; much heat evolves when added to water; very dangerous to eyes; wear face and eye protection when using this substance. Wear gloves. Hazard Code: B—Hazardous.

    2.5×10^{-5} M crystal violet: 0.020 g of crystal violet per 2 L of solution. **Note:** If a milligram balance is unavailable to you, prepare a 2.5×10^{-4} M crystal violet solution using 0.20 g per 2 L of solution. Then dilute 100 mL of this solution to a total volume of 1 L. Hazard Code: C—Somewhat hazardous. **Note:** Crystal violet leaves stains if spilled—be careful when preparing the solution. It also stains glassware if left for an extended time in glass containers. A solution of 1 M HCl will remove stains from glassware.

    The hazard information reference is: Flinn Scientific, Inc., *Chemical & Biological Catalog Reference Manual,* (800) 452-1261, www.flinnsci.com. See *Appendix D* of this book, *Chemistry with Vernier,* for more information.

3. The Logger *Pro* directions for this experiment, using a Vernier Colorimeter, instruct the student to open experiment "30b Rate Crystal Violet" from the Logger *Pro* experiment files. This experiment file is for a 200 second experiment length. **Note:** When using a Vernier or Ocean Optics Spectrometer, students do not use this file, but rather follow the directions in the written experiment for setting up the data collection.

    An alternative procedure for this experiment has been written in older editions of this lab book. In that version, a lower concentration of NaOH solution was used (0.02 M), the same concentration of crystal violet solution, and the same 10 mL volumes of both solutions. Using the lower concentration of NaOH caused the reaction to run slower, so that a 20 minute experiment was required. This experiment file is the experiment "30 Rate Crystal Violet" in Logger *Pro.* Unless you have a preference for the older method, there is no reason to use this file.

4. One extension of this experiment for advanced classes is to determine the order with respect to [OH$^-$]. This is done by performing the same procedure using a different concentration of OH$^-$. We recommend using 0.05 M NaOH. Combine equal quantities of 2.5×10^{-5} M crystal violet solution and 0.05 M NaOH. The rate constant found in this extension will be different than the one found in the main experiment. This is because these are *pseudo rate constants,* based on only one of the reactants (crystal violet). Once two different k values corresponding to two different OH$^-$ concentrations have been found, the order, n, for [OH$^-$] can be determined:

$$k_1 / k_2 = ([OH^-]_1 / [OH^-]_2)^n$$

5. There are two models of Vernier Colorimeters. The first model (rectangular shape) has three wavelength settings, and the newest model (a rounded shape) has four wavelength settings. The 565 nm wavelength of either model is used in this experiment. The newer model is an auto-ID sensor and supports automatic calibration (pressing the CAL button on the Colorimeter with a blank cuvette in the slot). If you have an older model Colorimeter, see www.vernier.com/til/1665.html for calibration information.

ANSWERS TO QUESTIONS

1. The graph of ln absorbance *vs.* time is linear, so the reaction is first order with respect to crystal violet concentration.

2. $k = -\text{slope} = -(-0.00617 \text{ s}^{-1}) = 0.00617 \text{ s}^{-1}$

3. rate = k[crystal violet]^1 = 0.00617[crystal violet]

4. A = 0.304 at 50 s, and A = 0.152 at 160 s, so the half-life = 160 – 50 = 110 s, or A = 0.244 at 82 s, and A = 0.122 at 196 s, so the half-life = 196 – 82 = 114 s.

    If students have used concentration-time formulas, they can also determine half-life by using the first-order formula: $\ln(A_0/A) = k \cdot t$:

    $$\ln(2/1) = (0.00617 \text{ s}^{-1}) \cdot t_{1/2}, \text{ therefore } t_{1/2} = 0.693/0.00617 \text{ s}^{-1} = 112 \text{ s}$$

SAMPLE DATA

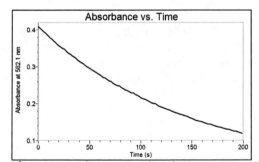

absorbance vs. time: reaction is not zero order

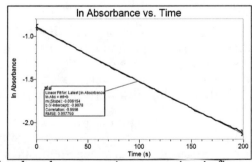

ln absorbance vs. time: reaction is first order

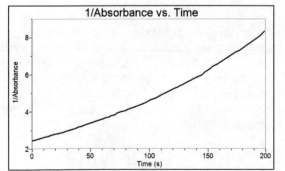

1/absorbance vs. time: reaction is not second order

Timed-Release Vitamin C Tablets

By slowly releasing a vitamin or a medicine throughout the day, timed-release tablets offer a simple alternative to multiple doses. There are several different methods used to achieve the timed-release effect. One common method involves covering the medicine with a polymer coating that allows it to slowly permeate into the body.

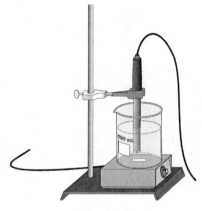

Figure 1

OBJECTIVES

In this experiment, you will

- Compare the behavior of timed-release vitamin C tablets with regular vitamin C tablets when each is added to distilled water.
- Use a pH Sensor to monitor the pH value of the two different types of vitamin C tablets over an elapsed time of approximately twelve minutes.

MATERIALS

computer	100 mL graduated cylinder
Vernier computer interface	one 500 mg regular vitamin C tablet
Logger *Pro*	one 500 mg timed-release vitamin C tablet
pH Sensor	distilled water
magnetic stirrer	wash bottle
stirring bar	ring stand
400 mL beaker	utility clamp

PROCEDURE

1. Obtain and wear goggles.

2. Measure out 250 mL of distilled water into a clean, dry 400 mL beaker.

3. Place the beaker onto a magnetic stirrer and add a small stirring bar.

4. Connect the pH Sensor to the computer interface. Prepare the computer for data collection by opening the file "31 Vitamin C" from the *Chemistry with Vernier* folder of Logger*Pro*.

5. Use a utility clamp to suspend a pH Sensor on a ring stand as shown in Figure 1. Place the pH Sensor in the beaker of water and adjust its position toward the outside of the beaker so that it is not struck by the stirring bar.

6. You are now ready to begin collecting data. Click ▶Collect to begin data collection and wait until the first point is recorded. This will be the initial pH of the distilled water. Add a 500 mg *timed-release* vitamin C tablet to the distilled water and set the magnetic stirrer at a brisk pace, so that a vortex is present with no splashing. **Important:** The initial pH reading should be between pH 6 and 7. If it is not, consult your teacher.

7. Continue with the experiment until data collection has stopped after 12 minutes. Store your data by choosing Store Latest Run from the Experiment menu. Dispose of the beaker contents as directed by your instructor.

8. Rinse the pH Sensor. After cleaning and drying the beaker, prepare a second 250 mL sample of distilled water. Click ▶Collect to begin data collection and wait until the first point is recorded—the initial reading must be between pH 6 and 7. With the magnetic stirrer at a brisk pace, add one 500 mg *regular* vitamin C tablet to the distilled water.

9. Continue with the experiment until data collection has stopped after 12 minutes. Dispose of the beaker contents as directed by your teacher. Rinse the pH Sensor and return it to the pH storage solution.

10. Both pH runs should now be displayed on the same graph. The plot for Run 1 (timed-release vitamin C) is red and the plot for Latest Run (regular vitamin C) is blue.

11. Find the pH of each type of vitamin-C tablet at 0, 4, 8, and 12 minutes.
    a. Click on the Examine button, ⬚.
    b. Examine the data pairs on the displayed graph. Based on this data, determine the pH value at 0, 4, 8, and 12 minutes for the first run (timed-released vitamin C) and the second run (regular vitamin C). Record these values in your data table

12. (optional) Print a graph of pH *vs.* time (with two curves displayed). Annotate each curve as *timed-release vitamin C* or *regular vitamin C*.

PROCESSING THE DATA

1. Using the pH values in the data table, compare the change in pH (ΔpH) during these three time intervals: 0–4 minutes, 4–8 minutes, and 8–12 minutes. Which vitamin-C tablet had the largest decrease in pH during each of these intervals? Explain.

2. Would you expect *one* 1000 mg vitamin C tablet or *two* 500 mg tablets to take more time to release the vitamin C (assume both are regular tablets)? Explain.

3. During the experiment, did you observe any differences in appearance or behavior of the two types of tablets?

4. What are some factors that could cause the results of this experiment to differ from what actually occurs when these tablets dissolve in your stomach?

5. (Optional) One tablet contains, according to the product description, a total of 500 mg of ascorbic acid ($H_2C_6H_6O_6$). If you were to dissolve 500 mg of ascorbic acid in 250 mL of solution, what would be the pH of the solution? How does this calculated pH compare to the final pH reading that appears on your titration curves? Which product is closer to reaching the calculated pH?

DATA TABLE

Time (min)	pH of timed-release vitamin C	pH of regular vitamin C
0		
4		
8		
12		

Timed-Release Vitamin C Tablets

1. The student pages with complete instructions for data-collection using LabQuest App, Logger *Pro* (computers), EasyData or DataMate (calculators), and DataPro (Palm handhelds) can be found on the CD that accompanies this book. See *Appendix A* for more information.

2. Prepare a large volume of fresh distilled water one or two days prior to the experiment. The water should have a pH value between 6 and 7. If your distilled water has been exposed to carbon dioxide in the air for a long period of time, the pH may drop below 6.

3. This lab activity is not meant to be a quantitative determination of the vitamin-C concentration of a tablet. Rather, your students should be directed to examine the data for trends that compare the manner in which the release of substances into a liquid system can be controlled.

4. The pH calibration that is stored in the data-collection software works well for this experiment. For more accurate pH readings, you (or your students) can do a 2 point calibration for each pH system using pH 4 and pH 7 buffers. See the Teacher Information in Experiment 25 for more detailed information about pH calibration.

5. Test results will vary somewhat, depending upon the brand of timed-release vitamin C that you use. You may choose to offer several types of products, or vary the size of tablets, to provide some interesting group data.

6. Most timed-release vitamin C tablets operate by diffusion and erosion of the tablet. A high-molecular-weight polymer (hydroxypropylmethylcellulose) is added to the tablet ingredients. As the polymer hydrates in the stomach (or beaker), it forms a gelatinous membrane that prevents the tablet from breaking apart quickly. This gel layer also controls the movement of water into the tablet and slows the migration of ascorbic acid out of the tablet. At the end of the testing, your students should observe that the regular tablet has broken apart and the timed-release tablet (while smaller) has retained its original shape. The timed-release tablet can be removed from the beaker and examined to reveal an essentially dry core.

SAMPLE RESULTS

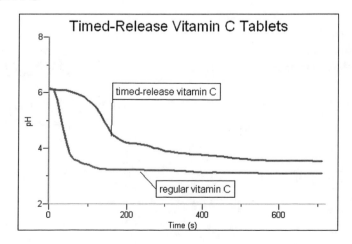

DATA TABLE

Time (min)	pH of timed-release vitamin C	pH of regular vitamin C
0	6.11	6.15
4	4.15	3.22
8	3.65	3.13
12	3.54	3.09

ANSWERS TO QUESTIONS

2. Between 0 and 4 minutes, the pH of the regular vitamin C had the biggest drop in pH (2.93 pH units, compared to 1.96 for the timed-release). Between 8 and 12 minutes, the timed-release vitamin C had a larger drop in pH (0.50 compared to 0.09 pH units for the regular vitamin C. Between 8 and 12 minutes, the timed-release vitamin C also had the biggest drop in pH (0.11 compared to 0.04 pH units for the regular. The test results will vary, depending on the brand of vitamin C used, but in all cases the regular vitamin C will have a faster initial decrease in pH, as expected. The timed-release tablet will have a more gradual release of vitamin C (although a surprisingly large percentage of the vitamin C is delivered in the first 12 minutes of the experiment).

3. This type of controlled-release method (polymer gel layer) is dependent upon total surface area for its release profile. The tablet with a smaller surface area per gram of vitamin C will release the substance at a slower rate. Generally, one 1000 mg tablet will have a smaller surface area than two 500 mg tablets.

4. The regular vitamin C tablet was completely broken apart at the end of the experiment and appeared to gradually disintegrate as the experiment proceeded. The timed-release tablet maintained its original shape and seemed to form a gel-like coating that was not present in the regular tablet.

5. In the stomach, the tablet is being added to dilute hydrochloric acid instead of water. The temperature of the stomach is higher than room temperature. The tablets would encounter less motion or turbulence in the stomach.

6. (Optional) A solution containing 0.500 gram of ascorbic acid in 250 mL of solution has a concentration of 0.0102 M. Using the K_{a1} of ascorbic acid as 7.9×10^{-5}, the pH of the solution is calculated to be 3.05. The results will vary depending on brands of tablets used, but in the sample data the regular vitamin C tablets had a final pH that was quite close to the calculated value. The calculations are shown here:

    molarity of ascorbic acid = [(0.500 g)(1 mol/176 g)] / (0.250 L) = 0.0102 M

    $K_a = 7.9 \times 10^{-5} = [H^+]^2 / (0.0102 \text{ M})$

    $[H^+] = 8.98 \times 10^{-4} \text{ M}$

    $pH = -\log[H^+] = -\log(8.98 \times 10^{-4})$

    $pH = 3.05$

The Buffer in Lemonade

One important property of weak acids and weak bases is their ability to form buffers. A buffer is the combination of a weak acid and a salt of the weak acid. Acetic acid and sodium acetate are an example of this kind of buffer pair. Buffers resist changes in pH upon the addition of small amounts of H^+ or OH^- ions. The dissociation equation for acetic acid contains both of the buffer components, $HC_2H_3O_2$ and $C_2H_3O_2^-$:

$$HC_2H_3O_2(aq) \longleftrightarrow H^+(aq) + C_2H_3O_2^-(aq)$$

When a small amount of an HCl solution is added to the buffer solution, most of the H^+ ions are removed when they react with acetate ions:

$$H^+(aq) + C_2H_3O_2^-(aq) \longleftrightarrow HC_2H_3O_2(aq)$$

When a solution of NaOH is added to the buffer, most of the OH^- ions are removed when they react with acetic acid molecules:

$$OH^-(aq) + HC_2H_3O_2(aq) \longleftrightarrow H_2O(l) + C_2H_3O_2^-(aq)$$

Buffers are incorporated into various consumer products to help control the effects of varying pH. The popular powdered drink mix used in this experiment uses a citric acid–sodium citrate buffer to "control acidity," according to its label.

OBJECTIVES

In this experiment, you will

- Use a pH Sensor to monitor pH as you titrate a given volume of the commercial brand of lemonade drink.
- Use a pH Sensor to monitor pH as you titrate an unbuffered solution of 0.010 M citric acid.
- Compare the results of the unbuffered solution with the lemonade buffer system.

MATERIALS

computer	distilled water
Vernier computer interface	50 mL or 100 mL graduated cylinder
Logger *Pro*	250 mL beaker
Vernier pH Sensor	ring stand
0.01 M citric acid solution	2 utility clamps
0.10 M NaOH solution	magnetic stirrer with stirring bar (if available)
lemonade drink	wash bottle with distilled water

PROCEDURE

1. Obtain and wear goggles.

2. Use a graduated cylinder to measure out 40 mL of the lemonade drink and 60 mL of distilled water into a 250 mL beaker. **CAUTION:** *Do not eat or drink in the laboratory.*

3. Place the beaker on a magnetic stirrer and add a stirring bar. If no magnetic stirrer is available, you will need to stir with a stirring rod during the titration.

4. Connect the pH Sensor to the computer interface. Prepare the computer for data collection by opening the file "32 Buffer Lemonade" from the *Chemistry with Vernier* folder of Logger*Pro*.

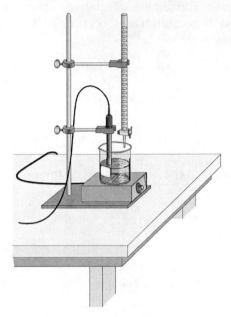

Figure 1

5. Use a utility clamp to suspend a pH Sensor on a ring stand as shown in Figure 1. Position the pH Sensor in the lemonade mixture and adjust its position so that it is not struck by the stirring bar.

6. Obtain a 50 mL buret and rinse the buret with a few mL of the 0.10 M NaOH solution. **CAUTION:** *Sodium hydroxide solution is caustic. Avoid spilling it on your skin or clothing.* Dispose of the rinse solution as directed by your teacher. Use a utility clamp to attach the buret to the ring stand as shown in Figure 1. Fill the buret a little above the 0.00 mL level of the buret with 0.10 M NaOH solution. Drain a small amount of NaOH solution so it fills the buret tip *and* leaves the NaOH at the 0.00 mL level of the buret. Record the precise concentration of the NaOH solution in your data table.

7. You are now ready to perform the titration. This process goes faster if one person manipulates and reads the buret while another person operates the handheld and enters volumes.

    a. Click ▶Collect to begin data collection.

    b. Before you have added any NaOH solution, click ⊕Keep and enter **0** as the buret volume in mL. Click ⬛OK⬛ to store the first data pair for this experiment.

    c. Add 2.0 mL of NaOH titrant. When the pH stabilizes, click ⊕Keep and enter the current buret reading in mL. Click ⬛OK⬛.You have now saved the second data pair for the experiment.

    d. Continue to add 2.0 mL increments, entering the buret level after each increment. When the pH has leveled off between 10.5 and 11, click ⬛Stop to end data collection. Dispose of the beaker contents as directed by your teacher.

8. Examine the data on the displayed graph to find the *equivalence point*—that is the largest increase in pH upon the addition of 2.0 mL of NaOH solution. As you move the examine line, the pH and volume values of each data point are displayed to the right of the graph. Go to the region of the graph with the largest increase in pH. Find the NaOH volume just *before* this jump. Record this value in the data table. Then record the NaOH volume *after* the 2.0 mL addition producing the largest pH increase.

9. Store the data from the first run by choosing Store Latest Run from the Experiment menu.

10. Use a graduated cylinder to measure out 40 mL of 0.010 M citric acid solution and 60 mL of distilled water into a 250 mL beaker. Position the pH Sensor, beaker, and stirring bar as you did in Step 5. Refill the buret to the 0.00 mL level of the buret with 0.10 M NaOH solution. **CAUTION:** *Sodium hydroxide solution is caustic. Avoid spilling it on your skin or clothing.*

11. Repeat Steps 7–8 of the procedure. Run 1 (lemonade drink) is plotted in red and Latest Run (unbuffered citric acid solution) is plotted in blue.

12. When you are finished, dispose of the beaker contents as directed by your teacher. Rinse the pH Sensor and return it to the pH storage solution.

13. Print a graph of pH *vs.* volume (with two curves displayed). Annotate each curve as *buffered lemonade* or *unbuffered citric acid*.

PROCESSING THE DATA

1. Determine the volume of NaOH added at the equivalence point for each trial. To do this, add the two NaOH values determined in Step 10 and divide by two.

2. Calculate the number of moles of NaOH used in each titration. Which solution reacted with more NaOH when the equivalence point was reached?

3. Examine the graph of each titration. How does the titration curve of the buffered lemonade compare to the curve of the unbuffered citric acid solution?

DATA TABLE AND CALCULATIONS

	Lemonade	Citric Acid solution
Concentration of NaOH	M	M
NaOH volume added *before* the largest pH increase	mL	mL
NaOH volume added *after* the largest pH increase	mL	mL
Volume of NaOH added at the equivalence point	mL	mL
Moles NaOH	mol	mol

TEACHER INFORMATION

The Buffer in Lemonade

1. The student pages with complete instructions for data-collection using LabQuest App, Logger *Pro* (computers), EasyData or DataMate (calculators), and DataPro (Palm handhelds) can be found on the CD that accompanies this book. See *Appendix A* for more information.

2. A two-quart volume of the lemonade drink, using tap water, should be prepared for the students.

3. Prepare the 0.010 M citric acid solution by dissolving 1.92 g of $H_3C_6H_5O_7$ (or 2.10 g of $H_3C_6H_5O_7 \cdot H_2O$) in 1.00 liter of solution. Hazard Code: D—Relatively non-hazardous.

4. Prepare 0.10 M NaOH solution by dissolving 4.0 grams of sodium hydroxide in 1.00 liter of solution. **HAZARD ALERT:** Corrosive solid; skin burns are possible; much heat evolves when added to water; very dangerous to eyes; wear face and eye protection when using this substance. Wear gloves. Hazard Code: B—Hazardous.

   The hazard information reference is: Flinn Scientific, Inc., *Chemical & Biological Catalog Reference Manual, 2000*, P.O. Box 219, Batavia, IL 60510. See *Appendix D* of this book, *Chemistry with Vernier*, for more information.

5. There are several brands of drink mix that use a citric acid–sodium citrate buffer. The sample data is based on Country Time Lemonade Mix, which works quite well. An interesting extension of this experiment would be to compare various brands for their effectiveness as buffers.

6. You may wish to have your students prepare their own citric acid–sodium citrate buffer solution. Students can start out with a 1:1 citric acid–sodium citrate ratio, and vary the ratio or the concentrations of the substances. Acid and salt concentrations of 0.01 M behave similarly to most commercial drink mixes when titrated with 0.1 M NaOH.

7. Adding a universal indicator at the beginning of each titration can provide a visual display of the activity of a buffer. The color changes can be quite interesting at the point where the buffer begins to fail.

8. The pH calibration that is stored in the data-collection software works well for this experiment. For more accurate pH readings, you (or your students) can do a 2-point calibration for each pH system using pH 4 and pH 7 buffers. See the Teacher Information in Experiment 25 for more detailed information about pH calibration.

SAMPLE RESULTS

Country Time Brand Lemonade Drink		0.01 M Citric Acid (unbuffered)	
pH	NaOH Vol. (mL)	pH	NaOH Vol. (mL)
2.835	0.0	2.791	0.0
3.007	2.0	3.094	2.0
3.202	4.0	3.720	4.0
3.396	6.0	4.432	6.0
3.590	8.0	5.144	8.0
3.784	10.0	5.921	10.0
3.978	12.0	7.281	12.0
4.151	14.0	10.842	14.0
4.324	16.0	11.101	16.0
4.518	18.0	11.230	18.0
4.691	20.0	11.295	20.0
4.907	22.0	11.338	22.0
5.144	24.0	11.381	24.0
5.338	26.0	11.403	26.0
5.597	28.0	11.446	28.0
5.835	30.0	11.489	30.0
6.115	2.0	11.532	32.0
6.482	34.0	11.554	34.0
7.324	36.0	11.597	36.0
10.151	38.0	11.619	38.0
10.518	40.0	11.640	40.0
10.712	42.0	11.684	42.0
10.842	44.0	11.705	44.0

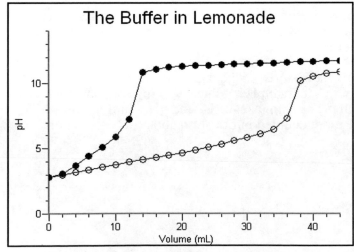

Titration of Country Time Lemonade drink and 0.01 M citric acid using 0.10 M NaOH

SAMPLE DATA AND CALCULATIONS

	Lemonade	Citric Acid
Concentration of NaOH	0.10 M	0.10 M
NaOH volume added before largest pH increase	36.0 mL	12.0 mL
NaOH volume added after largest pH increase	38.0 mL	14.0 mL
Volume of NaOH added at equivalence point	$\dfrac{36.00 + 38.00}{2} = 37.00$ mL 37.0 mL	$\dfrac{12.00 + 14.00}{2} = 13.00$ mL 13.0 mL
Moles NaOH	$(0.100 \text{ mol/L})(0.0370 \text{ L}) =$ 0.00370 mol	$(0.100 \text{ mol/L})(0.0130 \text{ L}) =$ 0.00130 mol

ANSWERS TO QUESTIONS

3. The lemonade drink neutralized nearly three times as much 0.10 M NaOH as did citric acid.

4. The graph of the citric acid titration shows a much steeper rise in pH as the base is added, and it requires much less of the base to neutralize the unbuffered citric acid.

Determining the Free Chlorine Content of Swimming Pool Water

Physicians in the nineteenth century used chlorine water as a disinfectant. Upon the discovery that certain diseases were transmitted by water, it became common for municipalities to chlorinate public water supplies. It is now standard practice to add chlorine to swimming pools and hot tubs. Chlorine reacts with water to form hypochlorous acid (HOCl) and hypochlorite ion (ClO⁻).

$$Cl_2 + H_2O \longrightarrow HOCl + H^+ + Cl^-$$

$$HOCl \longleftrightarrow H^+ + OCl^-$$

The chlorine that exists in water as HOCl and OCl⁻ is known as *free chlorine*. Free chlorine can kill bacteria, prevent algae growth, and oxidize iron to form a precipitate that can be removed from a pool by the filtering system. Swimming pool operators try to maintain a desired range of 1.0 to 1.5 mg/L of free chlorine for proper sanitation.

In this experiment, you will use a Colorimeter or Spectrometer to determine the amount of free chlorine in a sample of swimming pool or hot tub water. The solution used in this experiment has a red color, so Colorimeter users will be instructed to use the 565 nm (green) LED. Spectrometer users will determine an appropriate wavelength based on the absorbance spectrum of the solution. You will first measure the absorbance of light by aqueous solutions of known chlorine concentration. From the resulting graph of

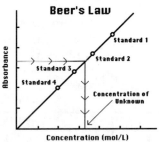

Figure 1

absorbance *vs.* free chlorine concentration (Beer's law), you will be able to determine the free chlorine content of your swimming pool sample.

A compound called DPD (N, N-diethyl-p-phenylenediamine) is reacted with the chlorine in each sample. The DPD is oxidized, forming a magenta (red) product. The intensity of the sample's color is directly proportional to its chlorine concentration.

OBJECTIVES

In this experiment, you will

- Prepare free chlorine standard solutions.
- Use a Colorimeter or Spectrometer to measure the absorbance of each standard solution.
- Plot a graph of absorbance of free chlorine *vs.* concentration.
- Use your Beer's law plot to determine the amount of free chlorine in a sample of swimming pool or hot tub water.

MATERIALS

computer	free-chlorine standard (10 mg/L)
Vernier computer interface*	water samples from a pool or hot tub
Logger *Pro*	six DPD free-chlorine powder pillows
Vernier Colorimeter or Spectrometer	25 mL pipet
one cuvette	10 mL pipet
tissues (preferably lint-free)	pipet pump or pipet bulb
six 50 mL beakers	stirring rod
100 mL beaker	distilled water

* No interface is required if using a Spectrometer

PROCEDURE

Both Colorimeter and Spectrometer Users

1. Obtain and wear goggles.

2. Label six clean, dry, 50 mL beakers 1–6. Transfer about 20 mL of the free-chlorine standard solution (10 mg/L) to a 100 mL beaker. **CAUTION:** *Handle this solution with care. Avoid breathing its vapors.* Use a 10 mL pipet to measure out exactly 1.00 mL of the free-chlorine standard solution into Beaker 1. Use a 25 mL pipet to add 24.00 mL of distilled water to Beaker 1. Continue using this procedure to prepare the free-chlorine solutions in Beakers 2–5, using the amounts of the free-chlorine standard solution and distilled water shown here:

Beaker number	Free-chlorine standard (10 mg/L) (mL)	Distilled water (mL)	Free-chlorine concentration (mg/L)
1	1.00	24.00	0.40
2	2.00	23.00	0.80
3	3.00	22.00	1.20
4	4.00	21.00	1.60
5	5.00	20.00	2.00

3. Use a 25 mL pipet to transfer 25.00 mL of your swimming pool water sample to Beaker 6.

4. Add one DPD free-chlorine powder pillow to each of the six labeled beakers. **CAUTION:** *Handle the powder pillows with care. Mix the powder thoroughly into the sample using a stirring rod.* Note: It is not important if a few small bits of powder do not dissolve.

5. Prepare a *blank* by filling an empty cuvette 3/4 full with distilled water. Seal the cuvette with a lid. To correctly use cuvettes, remember:

   - All cuvettes should be wiped clean and dry on the outside with a tissue.
   - Handle cuvettes only by the top edge of the ribbed sides.
   - All solutions should be free of bubbles.
   - Always position the cuvette so the light passes through the clear sides.

Spectrometer Users Only (Colorimeter users proceed to the Colorimeter section)

6. Use a USB cable to connect the Spectrometer to the computer. Choose New from the File menu.

7. To calibrate the Spectrometer, place the blank cuvette into the cuvette slot of the Spectrometer, choose Calibrate ▶ Spectrometer from the Experiment menu. The calibration

dialog box will display the message: "Waiting 90 seconds for lamp to warm up." After 90 seconds, the message will change to "Warmup complete." Click [OK].

8. Determine the optimal wavelength for creating this standard curve.

   a. Remove the blank cuvette and pour out the water. Using the solution in Beaker 1, rinse the cuvette twice with ~1 mL amounts and then fill it 3/4 full. Wipe the outside with a tissue and place it in the Spectrometer.

   b. Click [▶ Collect]. The absorbance *vs.* wavelength spectrum will be displayed. Click [■ Stop].

   c. To set up the data collection mode and select a wavelength for analysis, click on the Configure Spectrometer Data Collection icon, 🔍.

   d. Click Abs *vs.* Concentration (under the Set Collection Mode). The wavelength of maximum absorbance (λ max) is automatically identified. Click [OK].

   e. Proceed to Step 9.

Colorimeter Users Only

6. Connect the Colorimeter to the computer interface. Prepare the computer for data collection by opening the file "33 Free Chlorine" from the *Chemistry with Vernier* folder of Logger*Pro*.

7. Open the Colorimeter lid, insert the blank, and close the lid.

8. Calibrate the Colorimeter and prepare to test the standard solutions.

   a. Press the < or > button on the Colorimeter to select a wavelength of 565 nm (Green).

   b. Press the CAL button until the red LED begins to flash and then release the CAL button. When the LED stops flashing, the calibration is complete.

   c. Empty the water from the blank cuvette. Use the solution in Beaker 1 to rinse the cuvette twice with ~1 mL amounts and then fill it 3/4 full. Wipe the outside with a tissue and place it in the Colorimeter.

Both Colorimeter and Spectrometer Users

9. You are now ready to collect absorbance-concentration data for the five standard solutions.

   a. Leave the cuvette, containing the Beaker 1 mixture, in the device (Colorimeter or Spectrometer). Close the lid on the Colorimeter. Click [▶ Collect].

   b. When the value displayed on the monitor has stabilized, click [⊛ Keep] and enter **0.40** as the concentration in mg/L. Click [OK]. The absorbance and concentration values have now been saved for the first solution.

   c. Discard the cuvette contents as directed by your instructor. Using the solution in Beaker 2, rinse the cuvette twice with ~1 mL amounts and then fill it 3/4 full. Place the cuvette in the device. Wait for the value to stabilize and click [⊛ Keep]. Enter **0.80** as the concentration in mg/L, and click [OK].

   d. Repeat Step c for Beaker 3 (1.2 mg/L), Beaker 4 (1.6 mg/L), and Beaker 5 (2.0 mg/L). When you have finished with the Beaker 5 solution, click [■ Stop]. **Note:** Wait until Step 11 to measure the absorbance of the swimming pool water (Beaker 6).

   e. Examine the data pairs on the absorbance *vs.* concentration graph. As you move the examine line, the absorbance and concentration values of each data point are displayed to the right of the graph. Write down the absorbance and concentration data values in your data table.

10. Examine the graph of absorbance *vs.* concentration. To see if the curve represents a direct relationship between these two variables, click the Linear Fit button, 📈. A best-fit linear

regression line will be shown for your five data points. This line should pass near or through the data points *and* the origin of the graph. (**Note**: Another option is to choose Curve Fit from the Analyze menu, and then select Proportional. The Proportional fit has a y-intercept value equal to 0; therefore, this regression line will always pass through the origin of the graph).

11. Determine the absorbance value of the swimming pool water.

    a. Obtain 10 mL of the swimming pool water to be tested.

    b. Rinse the cuvette twice with the swimming pool water and fill it about 3/4 full. Wipe the outside of the cuvette and place it into the device.

    c. Read the absorbance value displayed in the meter. (**Important**: The reading in the meter is **not** necessary to click ▶Collect to read the absorbance value.) When the displayed absorbance value has stabilized, record the absorbance value for Trial 6, the swimming pool sample, in your data table.

12. Discard the solutions as directed by your instructor. Proceed directly to Steps 1–2 of Processing the Data.

PROCESSING THE DATA

1. Use the following method to determine the concentration of the swimming pool water sample. With the linear regression curve still displayed on your graph, choose Interpolate from the Analyze menu. A vertical cursor now appears on the graph. The cursor's concentration and absorbance coordinates are displayed in the floating box. Move the cursor along the regression line until the absorbance value is approximately the same as the absorbance value you recorded in Step 11. The corresponding concentration value is the free chlorine concentration of your swimming pool sample, in mg/L.

2. Print a graph of absorbance *vs.* concentration, with a regression line and interpolated unknown concentration displayed. To keep the interpolated concentration value displayed, move the cursor straight up the vertical cursor line until the tool bar is reached. Enter your name(s) and the number of copies of the graph you want.

DATA AND CALCULATIONS

Trial	Concentration (mg/L)	Absorbance
1	0.40	
2	0.80	
3	1.20	
4	1.60	
5	2.00	
6	Swimming pool sample	

Concentration of swimming pool sample	mg/L

TEACHER INFORMATION

Determining the Free Chlorine Content of Swimming Pool Water

1. The student pages with complete instructions for data-collection using LabQuest App, Logger *Pro* (computers), EasyData or DataMate (calculators), and DataPro (Palm handhelds) can be found on the CD that accompanies this book. See *Appendix A* for more information.

2. DPD free chlorine powder pillows, for use with a 25 mL test sample, can be purchased from Hach Company (order number 14070-99). The Hach Company toll-free telephone number is (800) 227-4224. These pillows can be used to test for free chlorine in the range of 0 to 2 mg/L.

3. A free chlorine standard solution may be purchased from Hach Company (order number 14268-10). Dilute this standard to 10 mg/L free chlorine using the concentration provided with the standard. (The standard will be in the range of 50–75 mg/L.)

4. For consistent results, students should use a graduated pipet to measure the volumes of free chlorine solution and distilled water when they prepare their dilutions.

5. If your swimming pool or hot tub water samples give absorbance values that appear to have a concentration that is higher than 2.0 mg/L, have students dilute the samples by a known amount so they fall in the 0 to 2.0 mg/L range. For example, if they dilute the concentration to one-half by combining equal volumes of water and free, determined the free-chlorine concentration of this diluted sample, then multiple the concentration by two.

6. If your students are already familiar with the "parts per million" (ppm) in ion analysis, you may want to explain that ppm and mg/L are equivalent units.

7. Show your students the procedure for opening Hach reagent pillows. You may want to supply each lab station with a pair of finger-nail clippers to open the pillows. Provide a location for them to return the empty pillows.

8. As an alternative to determining the concentration of the swimming pool water by interpolating along the regression line, it is also possible to *calculate* the concentration. If you get a good Beer's law curve (a straight line that passes through the origin of the graph), then you can divide the absorbance value of the tablet solution by the slope of the line.

9. There are two models of Vernier Colorimeters. The first model (rectangular shape) has three wavelength settings, and the newest model (a rounded shape) has four wavelength settings. The newer model is an auto-ID sensor and supports automatic calibration (pressing the CAL button on the Colorimeter with a blank cuvette in the slot). If you have an older model Colorimeter, see www.vernier.com/til/1665.html for calibration information.

SAMPLE RESULTS

Beaker	Concentration (mg/L)	Absorbance
1	0.40	0.030
2	0.80	0.067
3	1.20	0.121
4	1.60	0.161
5	2.00	0.210
6	swimming pool sample	0.066

Concentration of swimming pool sample	
	0.75 mg/L

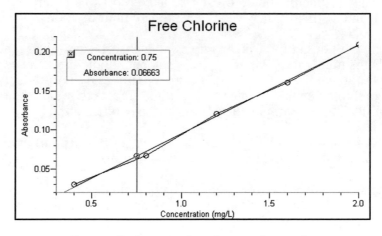

Interpolating a swimming pool sample

Determining the Quantity of Iron in a Vitamin Tablet

As biochemical research becomes more sophisticated, we are learning more about the role of metallic elements in the human body. For example, copper and zinc are present in enzymes, and trace amounts of molybdenum and selenium are vital in regulating internal oxidation-reduction reactions. Iron is necessary for oxygen transport in the bloodstream. Many people gain these essential elements through their diets, or by taking multivitamin tablets.

The iron that is present in these tablets is in the form of water-soluble Fe^{2+} ions. In this experiment, you will use the Colorimeter or Spectrometer to determine the quantity of iron in a vitamin tablet. You will prepare four solutions of known Fe^{2+} concentration. The Colorimeter or Spectrometer will be used to determine the absorbance of each solution at a specific wavelength of light. When a graph of absorbance *vs.* concentration is plotted for these solutions, a direct relationship should result (Beer's law).

You will also use a Colorimeter or Spectrometer to measure the absorbance of a solution prepared from a multivitamin tablet. This absorbance value is converted to concentration using the Beer's law curve. From this concentration, the total mass of iron in the original tablet can be calculated.

The iron in each tablet tested will be reacted with 1,10 phenanthroline so that it develops a detectable color. Sodium acetate and hydroxylamine hydrochloride are added to control the pH of the mixtures and keep the iron in the +2 oxidation state.

OBJECTIVES

In this experiment, you will

- Prepare Fe^{2+} standard solutions.
- Measure the absorbance of each standard solution.
- Plot a graph of absorbance of Fe^{2+} *vs.* concentration.
- Use your Beer's law plot to determine the quantity of iron in a vitamin tablet.

MATERIALS

computer	multivitamin tablet with iron
Vernier computer interface*	20 mg/L iron (II) solution
Logger *Pro*	1% hydroxylamine hydrochloride
Vernier Colorimeter or Spectrometer	1 M sodium acetate
one cuvette	1% 1,10 phenanthroline
four 20 x 150 mm test tubes	0.010 M HCl
test tube rack	distilled water
mortar and pestle	50 mL graduated cylinder
hot plate	two 100 mL beakers
filter funnel	two 10 mL graduated cylinders
filter paper	pipet pump or pipet bulb
stirring rod	two 10 mL pipets

* No interface is required if using a Spectrometer

PROCEDURE

Both Colorimeter and Spectrometer Users

1. Obtain and wear goggles.

2. Weigh one multivitamin tablet and record its mass to the nearest 0.001 g. Use a mortar and pestle to grind the tablet to a fine powder. Also record the mass of iron, in mg, listed on the label of the vitamin container.

3. Measure out 0.200 g of the powdered tablet and transfer it to a 100 mL beaker. **Note:** Try to avoid including bits of the tablet coating in your 0.200 g sample. Add 20 mL of 0.010 M HCl and 10 mL of distilled water. Mix thoroughly. **CAUTION:** *Handle the hydrochloric acid with care. It can cause painful burns if it comes in contact with the skin.*

4. Use a hot plate to heat the mixture in the beaker to a gentle boil. Boil the mixture for 5–7 minutes. Frequently stir the mixture with a stirring rod. After boiling, allow the mixture to cool to nearly room temperature. Some particles may remain undissolved, which is normal and will not affect your experiment.

5. Set up a funnel and filter paper. Pour the mixture through the filter, collecting the liquid filtrate in a 50 mL graduated cylinder. Rinse the 100 mL beaker with 10 mL of distilled water and pour the water through the filter. When the filter has drained, fill the graduated cylinder to the 50 mL mark with distilled water. Transfer this solution to a clean, dry, 100 mL beaker.

6. Label five clean, dry, test tubes 1–5. Prepare a set of four standards and one blank solution in the test tubes using the following solutions: **CAUTION:** *Handle the solutions with care.*

   A. 20 mg/L iron (II) solution
   B. 1% hydroxylamine hydrochloride
   C. 1 M sodium acetate
   D. 1% 1,10 phenanthroline

   Note: Solutions B, C, D, and the distilled water may be measured successively into one 10 mL graduated cylinder. Then add solution A, using a pipet for greater precision. Transfer the solution into a test tube. **Important:** Mix each solution thoroughly with a stirring rod.

Test Tube number	A (mL)	B (mL)	C (mL)	D (mL)	Distilled water (mL)
1	1.00	1.00	1.00	1.00	6.00
2	2.00	1.00	1.00	1.00	5.00
3	3.00	1.00	1.00	1.00	4.00
4	4.00	1.00	1.00	1.00	3.00
5 (blank)	0.00	1.00	1.00	1.00	7.00

7. Prepare a *blank* by filling an empty cuvette 3/4 full with distilled water. Seal the cuvette with a lid. To correctly use a cuvette, remember:

   • All cuvettes should be wiped clean and dry on the outside with a tissue.
   • Handle cuvettes only by the top edge of the ribbed sides.
   • All solutions should be free of bubbles.
   • Always position the cuvette so the light passes through the clear sides.

Spectrometer Users Only (Colorimeter users proceed to the Colorimeter section)

8. Use a USB cable to connect the Spectrometer to the computer. Choose New from the File menu.

9. To calibrate the Spectrometer, place the blank cuvette into the cuvette slot of the Spectrometer, choose Calibrate ▶ Spectrometer from the Experiment menu. The calibration dialog box will display the message: "Waiting 90 seconds for lamp to warm up." After 90 seconds, the message will change to "Warmup complete." Click [OK].

10. Determine the optimal wavelength for creating the standard curve.

    a. Remove the blank cuvette and pour out the water. Using the solution in Test Tube 1, rinse the cuvette twice with ~1 mL amounts and then fill it 3/4 full. Wipe the outside with a tissue and place it in the Spectrometer.

    b. Click [▶ Collect]. The absorbance *vs.* wavelength spectrum will be displayed. Click [■ Stop].

    c. To set up the data collection mode and select a wavelength for analysis, click on the Configure Spectrometer Data Collection icon, 🏠.

    d. Click Abs *vs.* Concentration (under the Set Collection Mode). The wavelength of maximum absorbance (λ max) is automatically identified. Click [OK].

    e. Proceed to Step 11.

Colorimeter Users Only

8. Connect the Colorimeter to the computer interface. Prepare the computer for data collection by opening the file "34 Iron in Vitamins" from the *Chemistry with Vernier* folder of Logger*Pro*.

9. Open the Colorimeter lid, insert the blank, and close the lid.

10. Calibrate the Colorimeter and prepare to test the standard solutions.

    a. Press the < or > button on the Colorimeter to select the wavelength of 565 nm (Green).

    b. Press the CAL button until the red LED begins to flash and then release the CAL button. When the LED stops flashing, the calibration is complete.

    c. Empty the water from the blank cuvette. Use the solution in Test Tube 1 to rinse the cuvette twice with ~1 mL amounts and then fill it 3/4 full. Wipe the outside with a tissue and place it in the Colorimeter.

Both Colorimeter and Spectrometer Users

11. You are now ready to collect absorbance-concentration data for the five standard solutions.

    a. Leave the cuvette, containing the Test Tube 1 mixture, in the device (Colorimeter or Spectrometer). Close the lid on the Colorimeter. Click [▶ Collect].

    b. When the value displayed on the monitor has stabilized, click [⊛ Keep] and enter **2.00** as the concentration in mg/L. Click [OK]. The absorbance and concentration values have now been saved for the first solution.

    c. Discard the cuvette contents as directed by your instructor. Using the solution in Test Tube 2, rinse the cuvette twice with ~1 mL amounts and then fill it 3/4 full. Place the cuvette in the device. Wait for the value to stabilize and click [⊛ Keep]. Enter **4.00** as the concentration in mg/L, and click [OK].

    d. Repeat Step 11c for Test Tube 3 (6.00 mg/L) and Test Tube 4 (8.00 mg/L). When you have finished with the Test Tube 4 solution, click [■ Stop].

   e. Examine the data pairs on the graph of absorbance *vs.* concentration. As you move the examine line, the absorbance and concentration values of each data point are displayed to the right of the graph. Write down the absorbance values in your data table.

12. Examine the graph of absorbance *vs.* concentration. To see if the curve represents a direct relationship between these two variables, click the Linear Fit button, ⬜. A best-fit linear regression line will be shown for your five data points. This line should pass near or through the data points *and* the origin of the graph. (**Note**: Another option is to choose Curve Fit from the Analyze menu, and then select Proportional. The Proportional fit has a y-intercept value equal to 0; therefore, this regression line will always pass through the origin of the graph).

13. Prepare the tablet solution in the 100 mL beaker for testing.

   a. Measure out 5.00 mL of the tablet solution into a second clean, dry, 100 mL beaker.

   b. Add 5.00 mL each of solutions B, C, and D, plus 30.0 mL of distilled water.

   c. Mix the solution thoroughly.

14. Find the absorbance of the tablet solution.

   a. Rinse the cuvette twice with the tablet solution you prepared in Step 13, and fill it 3/4 full.

   b. Wipe the outside of the cuvette and place it into the device. **CAUTION:** *Handle the solutions with care.*

   c. Read the absorbance value displayed in the meter. (**Important**: The reading in the meter is live, so it is **not** necessary to click ▶ Collect to read the absorbance value.) When the displayed absorbance value has stabilized, record the absorbance value for the tablet solution in your data table.

15. Discard the solutions as directed by your instructor. Proceed directly to Steps 1–4 of Processing the Data.

PROCESSING THE DATA

1. Use the following method to determine the iron concentration in tablet solution. With the linear regression curve still displayed on your graph, choose Interpolate from the Analyze menu. A vertical cursor now appears on the graph. The cursor's concentration and absorbance coordinates are displayed in the floating box. Move the cursor along the regression line until the absorbance value is approximately the same as the absorbance value you recorded in Step 14. The corresponding concentration value is the iron concentration of your tablet solution, in mg/L.

2. Print a graph of absorbance *vs.* concentration, with a regression line and interpolated unknown concentration displayed. To keep the interpolated concentration value displayed, move the cursor straight up the vertical cursor line until the tool bar is reached. Enter your name(s) and the number of copies of the graph you want.

3. Determine the mass of iron in your vitamin tablet.

    a. Because the tablet solution in Step 13 was diluted by a factor of 1/10, multiply your final iron concentration (in mg/L) by 10 to find the pre-dilution iron concentration, in mg/L.

    b. A portion of the tablet was dissolved in 50 mL of solution in Step 5. Use this volume and the answer in the previous step to calculate the mass of iron, in mg, in the tablet solution.

    c. Since the analysis used only a fraction of the total mass of the tablet (see Step 3), multiply the mg of iron in the previous step by the ratio of the total tablet mass to the 0.200 g portion:

    $$\text{mg of iron} \times \frac{\text{total tablet mass}}{0.200 \text{ g}} = \text{mg of iron in the vitamin tablet}$$

4. Compare your final answer in the previous step with the mg of iron shown on the label of vitamin tablets you tested. Use these values to calculate the percent error.

DATA AND CALCULATIONS

Mass of tablet	g
Mass of iron on label	mg

Test Tube Number	Concentration (mg/L)	Absorbance
1	2.00	
2	4.00	
3	6.00	
4	8.00	
tablet solution		

Iron concentration in the *pre-dilution* tablet solution (use a factor of 10)	mg/L
Mass of iron in the original tablet solution	mg
Calculated mass of iron in vitamin tablet	mg
Percent error	%

Determining the Quantity of Iron in a Vitamin Tablet

1. The student pages with complete instructions for data-collection using LabQuest App, Logger *Pro* (computers), EasyData or DataMate (calculators), and DataPro (Palm handhelds) can be found on the CD that accompanies this book. See *Appendix A* for more information.

2. We recommend having students do Steps 1–5 of the student procedure during one class period and Steps 6–13 during the following class period.

3. The wavelength setting for the iron (II) solution is the green LED (565 nm). The optimum wavelength for iron (II) is 508 nm, but good results can be obtained using the green LED.

4. A 100 mg/L iron (II) standard solution can be purchased from the Hach Company (order number 14175-42). The Hach Company toll-free number is (800) 227-4224.

   Add 20.0 mL of the 100 mg/L iron (II) standard solution to 80 mL of distilled water to obtain the 20.0 mg/L iron (II) standard solution used in this experiment.

5. If you elect to prepare your own 100 mg/L iron (II) standard solution, the following process is recommended:

   a. Slowly add 10 mL of concentrated sulfuric acid to 25 mL of distilled water. **HAZARD ALERT:** Severely corrosive to eyes, skin and other tissue; considerable heat of dilution with water; mixing with water may cause spraying and spattering. Solutions might best be made by immersing the mixing vessel in an ice bath. **Always add the acid to water, never the reverse**; extremely hazardous in contact with finely divided materials, carbides, chlorates, nitrites and other combustible materials. Hazard Code: A—Extremely hazardous.

   b. Dissolve 0.702 gram of ferrous ammonium sulfate (Mohr's salt), $Fe(NH_4)_2(SO_4)_2 \cdot 6H_2O$, in the sulfuric acid. Hazard Code: D—Relatively non-hazardous.

   c. Add 0.1 M $KMnO_4$ solution drop-by-drop until a faint pink color persists. **HAZARD ALERT:** Solution may be a skin irritant. Hazard Code: C—Somewhat hazardous.

   d. Dilute the solution to 1.00 liter.

   e. Add 20.0 mL of the 100 mg/L iron (II) standard solution to 80 mL of distilled water to obtain the 20.0 mg/L iron (II) standard solution used in this experiment.

6. The hydroxylamine hydrochloride ($H_2NOH \cdot HCl$) is 1% mass of solute to volume of solution. Prepare 500 mL of this solution by dissolving 5.0 g of the solid in 500 mL of distilled water.

   HAZARD ALERTS:

   Hydroxylamine hydrochloride (crystal): Toxic by ingestion; strong tissue irritant. Hazard Code: C—Somewhat hazardous.

7. The 1,10 phenanthroline ($C_{12}H_8N_2 \cdot H_2O$) is also 1% mass of solute to volume of solution. Prepare it by dissolving 5.0 g of the solid 1,10 phenanthroline in a warm (60°C) 40% ethanol-water solvent mixture (200 mL of ethanol and 300 mL of water).

HAZARD ALERTS:

1,10 phenanthroline (granular): Toxic. Hazard Code: C—Somewhat hazardous.

8. Prepare 0.010 M HCl by adding 100 mL of 0.10 M HCl per liter of solution. (0.10 M HCl requires 8.4 mL of concentrated HCl per liter of solution.) **HAZARD ALERT:** Highly toxic by ingestion or inhalation; severely corrosive to skin and eyes. Hazard Code: A—Extremely hazardous.

9. The 1.00 M sodium acetate solution is prepared by dissolving 8.20 g of sodium acetate ($NaC_2H_3O_2$) or 13.61 g of sodium acetate trihydrate ($NaC_2H_3O_2\cdot3H_2O$) in 100 mL of solution. **HAZARD ALERT:** Skin, eye and respiratory irritant. Hazard Code: D—Relatively non-hazardous.

The hazard information reference is: Flinn Scientific, Inc., *Chemical & Biological Catalog Reference Manual, 2000,* P.O. Box 219, Batavia, IL 60510. See *Appendix D* of this book, *Chemistry with Vernier,* for more information.

10. The sample data show the results of analysis of an Advanced Formula Centrum tablet. Iron is listed as 18 mg on the label. Any Centrum brand vitamin has sufficient iron for the experiment. This experiment can be expanded by offering a few different brands of vitamin supplement tablets. **Note:** Chewable vitamin tablets containing iron are not recommended for this experiment.

11. There are two models of Vernier Colorimeters. The first model (rectangular shape) has three wavelength settings, and the newest model (a rounded shape) has four wavelength settings. The newer model is an auto-ID sensor and supports automatic calibration (pressing the CAL button on the Colorimeter with a blank cuvette in the slot). If you have an older model Colorimeter, see www.vernier.com/til/1665.html for calibration information.

12. The cuvette must be from 55% to 100% full in order to get a valid absorbance reading. If students fill the cuvette 3/4 full, as described in the procedure, they should easily be in this range. To avoid spilling solution into the cuvette slot, remind students not to fill the cuvette to the brim.

SAMPLE RESULTS

Mass of tablet	1.302 g
Mass of iron on label	18.0 mg

Test tube number	Concentration (mg/L)	Absorbance
1	2.00	0.049
2	4.00	0.093
3	6.00	0.137
4	8.00	0.181
tablet solution	5.46	0.125

Iron concentration in the *pre-dilution* tablet solution (use a factor of 10)

5.46 mg/L x 10 = 54.6 mg/L

54.6 mg/L

Mass of iron in the original tablet solution

54.6 mg/L x 0.050 L = 2.73 mg

2.73 mg

Calculated mass of iron in vitamin tablet

2.73 mg x (1.302 g/0.200 g) = 17.8 mg

17.8 mg

Percent error

$$\frac{|18.0 - 17.8|}{|18.0|} \times 100 = 1.1\%$$

1.1 %

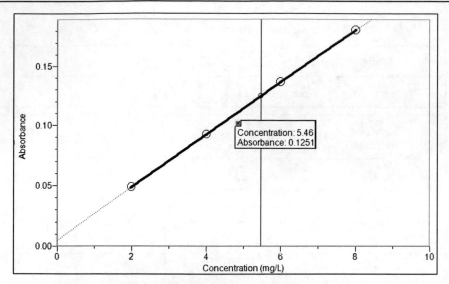

Absorbance vs. *iron concentration*

Determining the Phosphoric Acid Content in Soft Drinks

Phosphoric acid is one of several weak acids that exist in carbonated beverages. It is a component of all cola soft drinks. Phosphoric acid has a much higher concentration than other acids in a container of soft drink, so its concentration can be determined by a simple acid-base titration.

In this experiment, you will titrate a sample of a cola soft drink with sodium hydroxide solution and determine the concentration of phosphoric acid, H_3PO_4. Hydrogen ions from the first dissociation of phosphoric acid react with hydroxide ions from the NaOH in a one-to-one ratio in the overall reaction:

$$H_3PO_4(aq) + OH^-(aq) \longrightarrow H_2O(l) + H_2PO_4^-(aq)$$

In this experiment, you will use a pH Sensor to monitor pH as you titrate. The region of most rapid pH change will then be used to determine the equivalence point. The volume of NaOH titrant used at the equivalence point will be used to determine the molarity of the H_3PO_4.

OBJECTIVES

In this experiment, you will

- Use a pH Sensor to monitor pH during the titration of phosphoric acid in a cola soft drink.
- Using the titration equivalence point, determine the molarity of H_3PO_4.

MATERIALS

computer	0.050 M NaOH
Vernier computer interface	various cola soft drinks, decarbonated
Logger *Pro*	distilled water
Vernier pH Sensor	ring stand
50 mL buret	utility clamp
100 mL graduated cylinder	magnetic stirrer (if available)
250 mL beaker	stirring bar

PROCEDURE

1. Obtain and wear goggles.

2. Use a graduated cylinder to measure out 40 mL of a decarbonated cola beverage and 60 mL of distilled water into a 250 mL beaker. **CAUTION:** *Do not eat or drink in the laboratory.*

3. Place the beaker on a magnetic stirrer and add a stirring bar. If no magnetic stirrer is available, you need to stir with a stirring rod during the titration.

4. Connect the pH Sensor to the computer interface. Prepare the computer for data collection by opening the file "35 Phosphoric Acid" from the *Chemistry with Vernier* folder of Logger *Pro*.

5. Use a utility clamp to suspend a pH Sensor on a ring stand as shown in Figure 1. Position the pH Sensor in the HCl solution and adjust its position so that it is not struck by the stirring bar.

6. Obtain a 50 mL buret and rinse the buret with a few mL of the 0.050 M NaOH solution. **CAUTION:** *Sodium hydroxide solution is caustic. Avoid spilling it on your skin or clothing.* Dispose of the rinse solution as directed by your teacher. Use a utility clamp to attach the buret to the ring stand as shown in Figure 1. Fill the buret a little above the 0.00 mL level of the buret with 0.050 M NaOH solution. Drain a small amount of NaOH solution so it fills the buret tip *and* leaves the NaOH at the 0.00 mL level of the buret. Record the precise concentration of the NaOH solution in your data table.

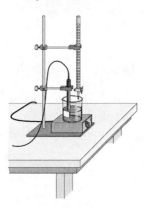

Figure 1

7. You are now ready to perform the titration. This process goes faster if one person manipulates and reads the buret while another person operates the handheld and enters volumes.

   a. Start data collection.

   b. Before you have added any NaOH solution, click ⎡⊛ Keep⎤ and enter **0** as the buret volume in mL. Click ⎡　OK　⎤ to store the first data pair for this experiment.

   c. Add 0.5 mL of NaOH solution. When the pH stabilizes, click ⎡⊛ Keep⎤ and enter the current buret reading in mL. Click ⎡　OK　⎤.You have now saved the second data pair for the experiment.

   d. Continue to add 0.5 mL increments, entering the buret level after each increment. When the pH has leveled off (near pH 10), click ⎡■ Stop⎤ to end data collection.

8. Examine the data on the graph of pH *vs.* volume to find the *equivalence point*—that is, the 0.5 mL volume increment that resulted in the largest increase in pH. As you move the examine line, the pH and volume values of each data point are displayed to the right of the graph. Go to the region of the graph with the large increase in pH. Find the NaOH volume (in mL) just *before* this jump. Record this value in the data table. Then record the NaOH volume *after* the 0.5 mL addition producing the largest pH increase.

9. Print a copy of the graph of pH *vs.* volume. Then print a copy of the NaOH volume data and the pH data for the titration.

10. Dispose of the beaker contents as directed by your teacher. Rinse the pH Sensor and return it to the storage solution.

PROCESSING THE DATA

1. Use your printed graph and data table to confirm the volume of NaOH titrant you recorded *before* and *after* the largest increase in pH values upon the addition of 0.5 mL of NaOH solution.

2. Determine the volume of NaOH added at the first equivalence point. To do this, add the two NaOH values determined above and divide by two.

3. Calculate the number of moles of NaOH used.

4. See the equation for the neutralization reaction given in the introduction. Determine the number of moles of H_3PO_4 reacted.

5. Recall that you pipeted out 40.0 mL of the beverage for the titration. Calculate the H_3PO_4 concentration.

DATA TABLE

Concentration of NaOH	M
NaOH volume added *before* the largest pH increase	mL
NaOH volume added *after* the largest pH increase	mL
Volume of NaOH added at equivalence point	mL
Moles NaOH	mol
Moles H_3PO_4	mol
Concentration of H_3PO_4	mol/L

Determining the Phosphoric Acid Content in Soft Drinks

1. The student pages with complete instructions for data-collection using LabQuest App, Logger *Pro* (computers), EasyData or DataMate (calculators), and DataPro (Palm handhelds) can be found on the CD that accompanies this book. See *Appendix A* for more information.

2. Decarbonate the cola soft drinks the day before doing the experiment. Boiling the beverages for 15–20 minutes will also remove most of the carbonation. If you boil the beverage, note the initial volume, then add distilled water to replenish evaporated liquid when you are finished.

3. The preparation of 0.050 M NaOH requires 2.0 g of NaOH per liter of solution. **HAZARD ALERT:** Corrosive solid; skin burns are possible; much heat evolves when added to water; very dangerous to eyes; wear face and eye protection when using this substance. Wear gloves. Hazard Code: B—Hazardous.

   The hazard information reference is: Flinn Scientific, Inc., *Chemical & Biological Catalog Reference Manual, 2000*, P.O. Box 219, Batavia, IL 60510. See *Appendix D* of this book, *Chemistry with Vernier,* for more information.

4. It is interesting to compare the acidity of several varieties of the same brand of cola soft drink, such as Pepsi, Diet Pepsi, Caffeine-Free Pepsi, Caffeine-Free Diet Pepsi, and Crystal Pepsi.

5. The pH calibration that is stored in the data-collection software works well for this experiment. For more accurate pH readings, you (or your students) can do a 2-point calibration for each pH system using pH 4 and pH 7 buffers. See the Teacher Information in Experiment 25 for more detailed information about pH calibration.

6. If your students have already done an acid-base titration, such as Experiment 24 or 25, you may want to have them use smaller NaOH volume increments near the equivalence point (e.g., 1 drop increments). Since the actual concentration of phosphoric acid will most likely be altered by the decarbonation process, we have written a procedure for determining only the *approximate* concentration of the acid. If you decide to use a more precise method, such as that in Experiment 24, you may want to use a lower NaOH concentration (e.g., 0.0200 M), so that a larger volume of NaOH is required to reach the first equivalence point.

7. A similar experiment from the May, 1983 edition of the Journal of Chemical Education (Joe Murphy, vol. 60, no. 5, pp. 420–1) compares titration to spectrophotometric analysis of phosphoric acid.

SAMPLE RESULTS

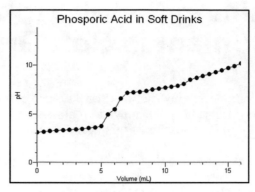

Titration of regular Coca Cola using 0.050 M NaOH

SAMPLE RESULTS

Concentration of NaOH	0.050 M
NaOH volume added before largest pH increase	5.00 mL
NaOH volume added after largest pH increase	5.50 mL
Volume of NaOH added at equivalence point	$\dfrac{5.00 + 5.50}{2} = 5.25$ mL 5.25 mL
Moles NaOH	(0.050 mol/L)(0.00525 L) = 0.000263 mol
Moles H_3PO_4	$\dfrac{1 \text{ mol } H_3PO_4}{1 \text{ mol NaOH}} \times 0.000263$ mol 0.000263 mol
Concentration of H_3PO_4	$\dfrac{0.000263 \text{ mol}}{0.0400 \text{ L}} =$ 0.0066 mol/L

Microscale Acid-Base Titration

A titration is a process used to determine the volume of a solution needed to react with a given amount of another substance. In this experiment, you will titrate hydrochloric acid solution, HCl, with a basic sodium hydroxide solution, NaOH. The concentration of the NaOH solution is given and you will determine the unknown concentration of the HCl. Hydrogen ions from the HCl react with hydroxide ions from the NaOH in a one-to-one ratio to produce water in the overall reaction:

$$H^+(aq) + Cl^-(aq) + Na^+(aq) + OH^-(aq) \longrightarrow H_2O(l) + Na^+(aq) + Cl^-(aq)$$

When HCl solution is titrated with NaOH solution, the pH value of the acidic solution is initially low. As base is added, the change in pH is quite gradual until close to the equivalence point, when equimolar amounts of acid and base have been mixed. Near the equivalence point, the pH increases very rapidly. The change in pH then becomes more gradual again, before leveling off with the addition of excess base.

Since this experiment may be your introduction to acid-base titrations, you will determine only the *approximate* concentration of the hydrochloric acid solution. Use the formula:

$$M_{acid} = M_{base} \times \frac{V_{base}}{V_{acid}}$$

where M_{acid} is the concentration of the acid (in M or mol/L), M_{base} is the concentration of the base, V_{base} is the volume of the base (in drops), and V_{acid} is the volume of the acid. The concentration of the sodium hydroxide solution is 0.10 M. The drops of sodium hydroxide and hydrochloric acid solutions at the equivalence point will be determined from the experiment.

OBJECTIVES

In this experiment, you will

- Perform a microscale acid-base titration.
- Monitor pH.
- Determine the approximate concentration of the acid used in the titration.

MATERIALS

computer	0.10 M NaOH solution (in a dropper bottle)
Vernier computer interface	HCl solution (in a dropper bottle)
Logger*Pro*	ring stand
Vernier pH Sensor	utility clamp
wash bottle	phenolphthalein indicator
distilled water	micro-beaker (top half of a storage bottle
toothpick (for stirring)	for the pH Sensor)

PROCEDURE

1. Obtain and wear goggles.

2. Connect the pH Sensor to the computer interface. Prepare the computer for data collection by opening the file "36 Microscale Titration" from the *Chemistry with Vernier* folder of Logger*Pro*.

3. Prepare the pH Sensor for data collection.

   a. Remove the pH Sensor from the pH storage solution bottle by unscrewing the lid. Carefully slide the lid from the sensor body.

   b. Rinse the tip of the sensor with distilled water.

   c. Obtain a pH Sensor storage bottle that has been cut in half. This is your microbeaker!

   d. With the open end of the pH Sensor pointing upward, as shown here, slip the microbeaker and cap down onto the sensor body (small opening first), so the sensor tip extends about 1 cm into the bowl of the microbeaker. Then tighten the threads of the cap so the cap tightens snugly against the pH Sensor body.

   e. Attach the utility clamp to a ring stand and to the bottle lid, with the sensor in an inverted position as shown here.

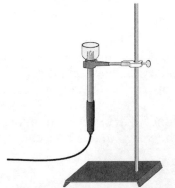

4. Obtain a dropper bottle containing the HCl solution of unknown concentration. Add 10 drops of the HCl solution into the micro-beaker. As you add the drops, hold the bottle in a vertical position to ensure that drop size is uniform. **CAUTION:** *Handle the hydrochloric acid with care. It can cause painful burns if it comes in contact with the skin.* Add 1 drop of phenolphthalein indicator to the microbeaker, then add enough distilled water so the resulting solution completely covers the sensor tip. Stir the solution thoroughly with the toothpick.

5. Obtain a dropper bottle containing 0.10 M NaOH. Wait until Step 7 to begin adding this solution to the HCl solution in the microbeaker.

6. Before adding NaOH titrant, click ▶Collect and monitor pH for 5-10 seconds. Once the pH has stabilized, click ⊛ Keep . In the edit box, type **0** (for 0 drops added). Press the ENTER key to store the first data pair for this experiment.

7. You are now ready to begin the titration. This process goes faster if one person adds the NaOH solution and stirs, while another person operates the computer and enters the number of drops.

   a. Add one drop of NaOH titrant to the micro-beaker. Be sure to hold the dropper bottle vertically to ensure that the drop size is uniform. **CAUTION:** *Sodium hydroxide solution is caustic. Avoid spilling it on your skin or clothing.* Stir with a toothpick to uniformly mix the solution. When the pH stabilizes, again click ⊛ Keep . In the edit box, type **1** for the number of drops of NaOH solution added. Press ENTER. You have now saved the second data pair for the experiment.

   b. Add a second drop of NaOH solution, stir, and click ⊛ Keep when the pH stabilizes. Enter **2** for the number of drops added.

   c. Continue this procedure until 20 drops of NaOH solution have been added.

8. When you have finished collecting data, click ⬛ **Stop** . Dispose of the beaker contents as directed by your teacher.

9. Print a copy of the graph. Enter your name(s) and the number of copies you want to print.

10. Print a copy of the table. Enter your name(s) and the number of copies you want to print.

PROCESSING THE DATA

1. Use your printed graph to confirm the volume of NaOH titrant you recorded *before* and *after* the largest increase in pH values upon the addition of 1 drop of NaOH solution.

2. Determine the volume of NaOH added at the equivalence point. To do this, add the two NaOH values determined above and divide by two (use 0.5-drop increments in your answer).

3. Using the formula in the introduction of the experiment, calculate the concentration of the hydrochloric acid solution (in M or mol/L).

DATA AND CALCULATIONS TABLE

Concentration of NaOH	M
NaOH volume added *before* the largest pH increase	drops
NaOH volume added *after* the largest pH increase	drops

Volume of NaOH added at equivalence point	
	drops
Concentration of HCl	
	M

Microscale Acid-Base Titration

1. The student pages with complete instructions for data-collection using LabQuest App, Logger *Pro* (computers), EasyData or DataMate (calculators), and DataPro (Palm handhelds) can be found on the CD that accompanies this book. See *Appendix A* for more information.

2. This experiment can be done prior to Experiment 24, "Acid-Base Titration". Students will quickly discover the shape of an acid-base titration curve for the reaction between a strong acid and strong base. They should not expect to determine precise concentration values using this method. It is meant to be an introduction to a traditional acid-base titration, not a substitute for more precise methods.

3. You can purchase additional pH storage solution bottles from Vernier to use as the microbeakers. Simply cut the bottles in half to use in this experiment. Distribute the half with the threaded opening to student lab stations. You do not need to include the cap, since students are instructed to use the cap and O-ring already on the probe). Order information is:

   pH Storage Solution Bottles   pkg of 5   order code: BTL

4. Explain to your students the purpose of adding the phenolphthalein indicator. They can easily observe the color change and large pH increase occur simultaneously in this experiment.

5. The preparation of 0.10 M NaOH requires 4.0 g of NaOH per liter of solution. Since the equivalence point concentrations are only approximate, using a value of ~0.10 M works well for this experiment. **HAZARD ALERT:** Corrosive solid; skin burns are possible; much heat evolves when added to water; very dangerous to eyes; wear face and eye protection when using this substance. Wear gloves. Hazard Code: B—Hazardous.

6. Unknown samples with HCl concentrations in the 0.080 to 0.100 M range work well. The preparation of 0.080 M HCl requires 6.7 mL of concentrated HCl per liter of solution. HCl that is 0.100 M requires 8.4 mL of concentrated reagent per liter. **HAZARD ALERT:** Highly toxic by ingestion or inhalation; severely corrosive to skin and eyes. Hazard Code: A— Extremely hazardous.

   The hazard information reference is: Flinn Scientific, Inc., *Chemical & Biological Catalog Reference Manual*, 1-800-452-1261, www.flinnsci.com. See *Appendix D,* of this book, *Chemistry with Vernier*, for more information.

7. The HCl and NaOH solutions can be dispensed from microscale Beral pipets if you do not have dropper bottles.

8. The stored pH calibration works well for this experiment.

SAMPLE RESULTS

Concentration of NaOH	0.10 M
NaOH volume added before largest pH increase	10 drops
NaOH volume added after largest pH increase	11 drops

Volume of NaOH added at equivalence point	$\dfrac{10 + 11}{2} =$ 10.5 drops
Concentration of HCl	$= 0.10 \text{ M} \times \dfrac{10.5 \text{ drops}}{10 \text{ drops}} =$ ~0.105 M

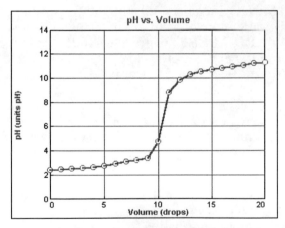

Microscale titration for sodium hydroxide and hydrochloric acid

Using the CD

The student pages with complete instructions for data-collection using LabQuest App, Logger *Pro* (computers), EasyData or DataMate (calculators), and DataPro (Palm handhelds) can be found on the CD that accompanies this book.

The CD located inside the back cover of this book contains the following.

📁 **Chemistry w Vernier Word**

> 📁 **Computer**–Supports Logger *Pro* 3.6 or newer computer software with LabPro and LabQuest. Contains the word-processing files for each of the 36 student experiments in this book

> 📁 **LabQuest**–Supports the LabQuest App with LabQuest. Contains files for each of the 36 student experiments in this book

> 📁 **Calculator**

>> 📁 **EasyData**–Supports the EasyData App with LabPro, CBL 2, or EasyLink. Contains files for each of the 36 student experiments in this book.

>> 📁 **DataMate**–Supports the DataMate calculator program with LabPro or CBL 2. Contains files for each of the 36 student experiments in this book.

> 📁 **Palm**–Supports the DataPro handheld program with LabPro. Contains the files for each of the student experiments in the book.

📁 **Word Files for Older Book**–Uses Logger *Pro* 2 computer software with LabPro, Serial Box Interface, or ULI. Contains the word-processing files for each of the 31 student experiments in the book, *Chemistry with Computers, Second Edition.*

📁 **Calculator Programs**

> 📁 **EasyData Calculator Program**–Contains the EasyData App for TI-83 Plus and TI-84 Plus graphing calculators.

> 📁 **TI Connect Software**–Contains the TI Connect software for Mac and Windows that is used to transfer programs and applications from a computer to a TI graphing calculator.

Using the *Chemistry with Vernier* Word-Processing Files

Start Microsoft Word, then open the file of your choice from the Chemistry w Vernier Word folder. Files can be opened directly from the CD or copied onto your hard drive first. These files can be used with any version of Microsoft Word that is Word 97 or newer.

All file names begin with the experiment number, followed by an abbreviation of the title; e.g., 01 Endo-Exothermic is the file name used for *Experiment 1, Endothermic and Exothermic Reactions.* This provides a way for you to edit the tests to match your lab situation, your equipment, or your style of teaching. The files contain all figures, text, and tables in the same format as printed in *Chemistry with Vernier.*

Using TI Connect: Loading EasyData and Capturing Calculator Screen Images

This appendix gives an overview of loading and updating the EasyData application on your TI-83 Plus or TI-84 Plus graphing calculators. It also describes how to download calculator screen images, such as graphs, to a computer.

EasyData is part of the bundle of APPS that come preloaded on all new TI-83 Plus, TI-84 Plus and TI-84 Plus Silver Edition graphing calculators. To do the activities in this book, you will need to use version 2.0 or newer. To check to see if you have the correct version of EasyData on your calculator, press ⌐APPS⌐ and scroll through the list of applications. With EasyData highlighted, press ⌐ENTER⌐ to launch the App and note the version number on the introductory screen. If your graphing calculator does not contain an appropriate version of EasyData, you can download the latest version for free from the Vernier web site www.vernier.com/easy.html and use TI Connect to transfer it to your graphing calculator.

Before loading EasyData onto your calculator, make sure you have a recent version of the calculator operating system. If you are using a TI-84 Plus calculator, you will need operating system version 2.30 or newer. If you are using a TI-83 Plus calculator, you will need operating system version 1.18 or newer. Operating system updates can be downloaded from the Texas Instruments web site at education.ti.com.

Loading EasyData

TI Connect for Windows and Macintosh is simple and easy to use. First, be sure you have the TI Connect software installed on your computer. If you do not, you can download this software for free from the Texas Instruments web site at www.education.ti.com/ticonnect. When you have downloaded the EasyData App onto your computer, follow the instructions below to transfer the EasyData app to your graphing calculator.

Windows Computers running Windows 98, NT 4.x, 2000 or ME, and XP

1. Connect the TI-GRAPH LINK cable or the TI Connectivity cable to the serial or USB port of your computer and to the port at the bottom edge of the TI-83 Plus graphing calculator.

    If you are using the TI-84 Plus or TI-84 Plus Silver Edition, connect the TI unit-to-computer cable to the USB port of your computer and to the USB port at the top edge of your graphing calculator.

2. Start the TI Connect software on your computer. Within TI Connect, click on Device Explorer. You may be asked to identify which computer port your cable is plugged into.

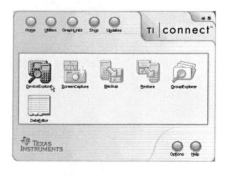

3. The software will identify the attached calculator and call up a window showing the calculator's contents.

4. To load EasyData onto a TI graphing calculator, simply drag the EasyData file from wherever you have it saved on your computer to the Device Explorer window, and it will copy onto your graphing calculator.

5. The EasyData App should now be loaded into your calculator. To confirm this, press ⟨APPS⟩ on the calculator to display the loaded applications.

Macintosh Computers running Mac OS X 10.2 (Jaguar), 10.3 (Panther), and 10.4 (Tiger).

1. Connect the TI-GRAPH LINK cable, or the TI Connectivity cable to the USB port of your computer and to the port at the bottom edge of the TI-83 Plus, or TI-83 Plus Silver Edition graphing calculator.

   If you are using the TI-84 Plus or TI-84 Plus Silver Edition graphing calculator, connect the TI unit-to-computer cable to the USB port of your computer and to the USB port at the top edge of your graphing calculator.

2. Turn the calculator on. On the computer, start TI Connect and double-click on TI Device Explorer.

3. The program will identify the attached calculator and call up a window displaying the calculator's contents.

4. To load EasyData onto a TI graphing calculator, simply drag the EasyData file from wherever you have it saved on your computer to the Device Explorer window, and it will copy onto your graphing calculator.

5. The EasyData App should now be loaded into your calculator. To confirm this, press ⟨APPS⟩ on the calculator to display the loaded applications.

Capturing a Calculator Screen Image

Many experiments in *Chemistry with Vernier* contain an optional step that involves printing the graph displayed on the calculator screen. To capture and print a TI-83 Plus calculator screen image, you need a TI-GRAPH LINK or TI Connectivity cable and the TI Connect software.

If you are using the TI-84 Plus or TI-84 Plus Silver Edition graphing calculator, connect the TI unit-to-computer cable to the USB port of your computer and to the USB port at the top edge of your graphing calculator.

Before doing an experiment that requires printing, you may want to show your students how to print graphs. The process described below produces screen images directly from the calculator, such as seen here.

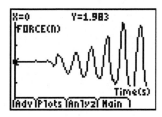

A graph of force vs. time as displayed in the EasyData application

Windows Computers running Windows 98, NT 4.x, 2000 or ME, and XP

1. On your calculator, display the screen you want to capture.

2. Connect the TI-GRAPH LINK cable, or the TI Connectivity cable to the serial or USB port of your computer and to the port at the bottom edge of the TI-83 Plus graphing calculator.

    If you are using the TI-84 Plus or TI-84 Plus Silver Edition, connect the TI unit-to-computer cable to the USB port of your computer and to the USB port at the top edge of your graphing calculator.

3. Start the TI Connect software on your computer. Within TI Connect, click on ScreenCapture.

    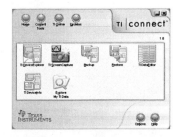

4. The captured screen will appear in TI Screen Capture window, and it can now be printed or saved to a file. To capture an additional screen, select Get Screen from the Tools menu.

    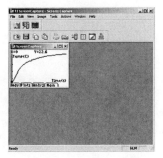

Macintosh Computers running Mac OS X 10.2 (Jaguar), 10.3 (Panther), and 10.4 (Tiger).

1. On your calculator, display the screen you want to capture.

2. Connect the TI-GRAPH LINK cable to the USB port of your computer and to the port at the bottom edge of the TI-83 Plus graphing calculator.

   If you are using the TI-84 Plus or TI-84 Plus Silver Edition, connect the TI unit-to-computer cable to the USB port of your computer and to the USB port at the top edge of your graphing calculator.

3. Start the TI Connect software on your computer. Within TI Connect, double-click on TI Device Explorer.

4. The TI Device Explorer window will appear displaying the contents of the calculator's memory.

5. Click the "Screen Capture" toolbar control. The calculator's screen image will appear in a separate window. The captured screen can now be printed or saved to a file.

Using Logger *Pro* to Transfer Data to a Computer

You may elect to use the Vernier Logger *Pro* program to transfer data from LabQuest, a TI graphing calculator, or Palm handheld to a computer. Logger *Pro* has many graphing features, such as labels and units for axes, autoscaling, and modification of axes. Printed graphs will have a better resolution and appearance than printed screens of the LabQuest, TI graphing calculator, or Palm handheld display. Data tables can be displayed and printed with side-by-side columns and headings. Logger *Pro* also provides advanced data-analysis features, such as curve fitting, statistical analysis, and calculated spreadsheet columns.

The directions below are for the latest version of Logger *Pro* (version 3.0 or newer).

Transferring data from LabQuest

Logger *Pro* can open files saved in the LabQuest application.

1. Connect the LabQuest to your computer with a USB cable.

2. Start Logger *Pro* on your computer.

3. Choose LabQuest Browser ▶ Open… from the File menu.

4. You'll see a standard file selection dialog showing the files available on your LabQuest. Select the file name you want, and click Open. Logger *Pro* will open the LabQuest file, displaying any data, graphs, and notes.

Transferring data from a TI graphing calculator

Logger *Pro* software has an option for importing data lists from all of the TI graphing calculators.

1. If you are using the TI-83 Plus or TI-83 Plus Silver Edition graphing calculator, connect the TI-GRAPH LINK cable or the TI Connectivity cable to the serial or USB port of your computer and to the port at the bottom edge of the calculator.

    If you are using the TI-84 Plus or TI-84 Plus Silver Edition, connect the TI unit-to-computer cable to the USB port of your computer and to the USB port at the top edge of your graphing calculator.

2. Turn on the calculator.

3. Start Logger *Pro* on your computer.

4. Choose Import from▶ TI Device from the File menu. A dialog box appears with directions for importing data.

5. From the pull-down menu, choose the USB port or serial port (COM 1-4 on a PC, modem or printer port on a Macintosh) to which the TI-GRAPH LINK cable is connected.[1]

---

[1] If you are using a PC serial cable, identify whether it is a gray or black cable.

6. Click on the Scan for Calculator button. The calculator model you are using should now be identified, and you should see a message, "Ready to Import."

7. Select the lists that you wish to import by clicking on each of them. (To select more than one list on a Macintosh, hold down the Command and Shift keys while you click.)

8. Click OK to send the lists to the computer. The lists will appear in columns in the data table in Logger *Pro*. They will be labeled with the simple list names from the calculator. If you want to rename them or add units, double-click on the heading in the data table and enter new labels and units.

9. Click the Refresh Catalog button if you have connected a new interface or calculator to the computer.

Transferring data from a Palm handheld

Now that you have collected data, you can, of course, analyze it with your Palm device. However, if you wish to analyze the data on a computer or print the data, you will need to use the Palm HotSync cradle or cable (or USB cable with Dana from AlphaSmart) to transfer the data back to your computer.

Vernier Logger *Pro* software (Versions 3.1 or newer for Macintosh or Windows) has a file option for importing data lists from all of the LabPro-compatible Palm devices. Logger *Pro* is the best option because it can also be used as a computer-based data collection program with LabPro. It is a great tool for *directly* moving LabPro data from the Palm to the computer for graphing and printing. Both are also very good for performing mathematical modeling and finding the relationships between variables. Or, if you do not own Logger *Pro*, we also include directions below for importing data into Microsoft Excel, or other spreadsheet programs.

There are two ways to import the most recent data from a Palm device using Logger *Pro*. The first method is the recommended method:

1. Use HotSync, with Logger *Pro* Already Running

a. Start Logger Pro on your PC or Macintosh computer.

b. Connect your Palm handheld to the HotSync cradle/cable. If you are using a Dana, connect the USB cable to the Dana's computer port (using the square or "B" end of the USB cable) and to the computer's USB port (using the flat or "A" end of the USB cable).

c. Press the HotSync button ⟳™ on the cradle/cable. On Dana, press the function and sync keys.

    d. You should get a message on your computer screen indicating that your Palm device and computer are communicating.

    e. The data will be imported into the data table object and displayed graphically, as shown in the screen above.

2. Select Import From→ Palm Data Pro... from the File Menu

    a. Connect your Palm handheld to the HotSync cradle/cable. If you are using a Dana, connect the USB cable to the Dana's computer port (using the square or "B" end of the USB cable) and to the computer's USB port (using the flat or "A" end of the USB cable).

    b. Press the HotSync button ⊙™ on the cradle/cable. On Dana, press the **function** and **sync** keys.

    c. You should get a message on your computer screen indicating that your Palm device and computer are communicating. When this process is completed, the data file will be in the Data Pro folder in the Palm User folder on your computer.

    d. Start Logger *Pro* on your PC or Macintosh computer.

    e. Select Import From► Palm Data Pro from the File Menu.

    f. The data will be imported into the data table and displayed graphically, as shown on the previous page.

Load Data from Files Saved by Data Pro Perform all the steps in number 2 above, but select Import From► Text File from the File Menu. Navigate to the Palm/User Folder/Data Pro folder and select the saved text file you wish to import.

Load Text Files into other Applications, such as Microsoft Excel The data that you send to the computer using HotSync can also be loaded into spreadsheet applications, such as Microsoft Excel. DataProX.txt file is found in Palm/User Folder/Data Pro.)

Use HotSync

    a. Connect your Palm handheld to the HotSync cradle/cable. If you are using a Dana, connect the USB cable to the Dana's computer port (using the square or "B" end of the USB cable) and to the computer's USB port (using the flat or "A" end of the USB cable).

    b. Press the HotSync button ⊙™ on the cradle/cable. On Dana, press the **function** and **sync** keys.

    c. You should get a message on your computer screen indicating that your Palm handheld and computer are communicating.

When this process is completed, the data file will be stored in the Data Pro folder in the Palm User folder on your computer, as a text file (DataProX.txt).

d. Start a spreadsheet application, such as Microsoft Excel. Navigate to the location of DataProX.txt (in Palm/User Folder/Data Pro) on your computer, and import the file into spreadsheet application.

Safety Information

Chemical Hazard Information

The reference source for the chemical hazard information in this book is the 2002 edition of Flinn Scientific's *Chemical & Biological Catalog Reference Manual*. Flinn Scientific, Inc. is an acknowledged leader in the areas of chemical supply, apparatus and laboratory equipment supply, and chemical safety. Flinn's *Chemical & Biological Catalog Reference Manual* is an outstanding reference to be used as you order chemicals, store chemicals, mix solutions, use chemicals in you classroom, and dispose of chemicals. Most of the chemicals and the equipment used in *Biology with Vernier* are available from this catalog. We strongly urge you to obtain and use a current copy of the above mentioned publication by contacting Flinn Scientific at the address below:

> Flinn Scientific, Inc.
> P.O. Box 219
> Batavia, Illinois 60510
> Telephone (800) 452-1261
> www.flinnsci.com

The Flinn hazard code is used in the teacher information section of many tests in *Biology with Computers* to describe any possible hazards associated with the chemical reagents used. The Flinn hazard code (A–D) is defined as follows:

A. Extremely Hazardous. This category includes, but is not limited to, concentrated acids, severely toxic, severely corrosive, unstable and /or explosive chemicals.

B. Hazardous. This category includes, but is not limited to, chemicals that are toxic/poisons, corrosive, contain heavy metals, and/or are alleged/proven carcinogens.

C. Somewhat Hazardous. This category includes, but is not limited to, chemicals that are highly flammable/combustible, moderately toxic and/or oxidants.

D. Relatively Non-Hazardous. This category includes, but is not limited to, chemicals that are irritants and/or allergens.

Vernier Products for Chemistry

The Vernier Software & Technology products required to perform the *Chemistry with Vernier* experiments are described in this appendix. A table of order codes can be found on the next page.

Interface Options

LabQuest

Vernier LabQuest provides a portable and versatile data-collection device for any class studying chemistry. It can be used as a computer interface, as a stand-alone device, or in the field. It has built-in graphing and analysis software and a vivid color touch screen. It is compatible with existing Vernier sensors. It has a rechargeable, high-capacity internal battery. It also has a built-in temperature sensor and microphone.

LabQuest Mini

The Vernier LabQuest Mini is a low-cost data-collection interface that connects to the USB port of a computer and has five sensor ports.

LabPro

Vernier LabPro offers another option for data collection in chemistry. A wide variety of Vernier probes and sensors can be connected to each of the four analog channels and two sonic/digital channels. LabPro is connected to a computer using a serial or USB port or to a TI graphing calculator or Palm handheld.

Go! Link

Go! Link is a single channel interface that plugs into the USB port of the computer. It is used with Logger *Pro* for data collection. Go! Link's flexibility and ease of use make it perfect for a variety of activities in science and math. It supports over 30 analog sensors, including Gas Pressure, pH, and Conductivity, among others.

CBL 2

Texas Instruments CBL 2 is also a portable and versatile data collection device. A wide variety of Vernier probes and sensors can be connected to each of the three analog channels and one digital/sonic channel of the CBL 2. The CBL 2 is connected to a TI calculator through the port found on the bottom edge of the calculator. Because the CBL 2 is battery powered, it can be taken out of the classroom to monitor data in the field. Within the classroom, the CBL 2 provides a low-cost alternative to the use of computers.

EasyLink

EasyLink is a single channel interface that plugs into the USB port of the TI-84 Plus or TI-84 Plus Silver Edition calculator. It is used with the EasyData App for data-collection. EasyLink's flexibility and ease of use make it perfect for a variety of activities in science and math. It supports over 30 analog sensors, including Gas Pressure, pH, and Conductivity, among others.

Data-Collection Software for Chemistry
Computer

Logger *Pro* is the data-collection software for collecting data on a computer. Logger *Pro* software comes with a free site license for both Windows and Macintosh, so you only need to order one copy of Logger *Pro* for your school or college department.

LabQuest

LabQuest App is the data-collection application used to collect data when using LabQuest as a standalone device. Included with LabQuest purchase.

Calculator

The EasyData App controls the data gathering process, and makes data analysis easier after experiments are completed. See *Appendix B* for information on transferring the program to your calculator.

Vernier Products for *Chemistry with Vernier*

Item	Order Code
Vernier LabQuest interface	LABQ
Vernier LabQuest Mini	LQ-MINI
Vernier LabPro interface	LABPRO
Go! Link interface	GO-LINK
EasyLink interface	EZ-LINK
Stainless Steel Temperature Probe	TMP-BTA
pH Sensor	PH-BTA
Gas Pressure Sensor	GPS-BTA
Conductivity Probe	CON-BTA
Colorimeter	COL-BTA
Drop Counter	VDC-BTD
Voltage Probe	included with LabPro and CBL 2
SpectroVis Plus	SVIS-PL
Chemistry with Vernier lab manual	CWV

Optional Vernier Products

Logger *Pro* (Windows *and* Macintosh)	LP
Replacement Cuvettes (package of 100)	CUV
pH Buffer Set (for pH 4, 7, and 10 buffers)	PHB
pH Storage Solution (500 mL bottle)	PH-SS
Stir Station	STIR
Dissolved Oxygen Probe	DO-BTA
Thermocouple	TCA-BTA

Vernier Sensors for Chemistry

Stainless Steel Temperature Probe

The Stainless Steel Temperature Probe is an accurate, durable, and inexpensive sensor for measuring temperature. Range: –25°C to +125°C

pH Sensor

Our pH Sensor is a Ag-AgCl gel-filled combination electrode and amplifier. It includes a convenient storage solution container that can be attached directly to the electrode. Range: 0 to 14 pH units

Gas Pressure Sensor

The Gas Pressure Sensor can be used for pressure-volume, pressure-temperature, and vapor pressure experiments in chemistry. The pressure range is 0 to 2.1 atm (0 to 210 kPa). It comes with a variety of pressure-sensor accessories, including a syringe, plastic tubing with two Luer-lock connectors, two rubber stoppers with Luer-lock adapters, and one two-way valve.

Colorimeter

The Vernier Colorimeter allows you to study the light absorption of various solutions. It is great for Beer's law experiments, determining the concentration of unknown solutions, or studying changes in concentration vs. time. Fifteen 3.5 mL cuvettes are included.

Conductivity Probe

This probe is great for environmental testing for salinity, total dissolved solids (TDS), or conductivity in water samples. Chemistry students can use it to investigate the difference between ionic and molecular compounds, strong and weak acids, salinity, or ionic compounds that yield different ratios of ions. The Conductivity Probe can monitor concentration or conductivity at three different sensitivity settings: 0–200 µS/cm, 0–2000 µS/cm, and 0–20,000 µS/cm.

Dissolved Oxygen Probe

Use the Dissolved Oxygen Probe to determine the concentration of oxygen in aqueous solutions in the range of 0–14 mg/L (ppm). It has built-in temperature compensation and a fast response time. This probe is great for water quality, chemistry, or ecology. Included with the probe is a zero-oxygen solution, two membrane caps, a 100% calibration bottle, and electrode filling solution. Replacement membranes are available (order code MEM).

Thermocouple

A Thermocouple sensor is used to measure temperatures over a wide range. The Vernier Thermocouple uses a type K thermocouple wire which can be used in the range of –200 to 1400°C.

Drop Counter

Use the Drop Counter with a pH Sensor, Conductivity Probe, or ORP Sensor to easily perform conductimetric or potentiometric titrations.

Equipment and Supplies

A list of equipment and supplies for all the experiments is given below. The amounts listed are for a class of up to 30 students working in groups of two, three, or four students in a classroom using eight calculators with interfaces. The materials have been divided into **nonconsumables**, **consumables**, and **chemicals**. Most consumables and chemicals will need to be replaced each year. Most nonconsumable materials may be used many years without replacement. Some substitutions can be made.

Nonconsumables

Item	Amount	Experiment
alcohol burner	8	17
alligator clips	16	29
balance	2	1, 19
balance, milligram	2	25
beaker, 50 mL	30	13, 21, 33
beaker, 100 mL	24	3, 8, 11, 14, 20, 22, 27, 33, 34
beaker, 250 mL	16	1, 4, 12, 13, 18, 19, 21, 23, 24, 25, 26, 29, 30, 32, 35
beaker, 400 mL	8	2, 12, 15, 31
beaker, 1 L	8	3, 7, 10
buret, 50 mL	8	23, 24, 25, 26, 35
can, small	8	16, 17
clamp, utility (buret clamp)	24	2, 3, 4, 7, 8, 12, 13, 14, 15, 16, 17, 19, 21, 22, 23, 24, 25, 26, 31, 32, 35, 36
clock (with second hand)	8	29
condenser with two hoses	8	8
cuvette	8	11, 20, 30, 33, 34
dropper bottle	16	36
electric motor, small	8	29
flask, 100 mL volumetric	8	27
flask, 125 mL Erlenmeyer	16	7, 10
flask, 500 mL	8	8
food holder	8	16
forceps	8	28
funnel	8	34

glass plate, 15 x 15 cm	8	28
glove, heat-proof	8	7
goggles, chemical splash	30	all
graduated cylinder, 10 mL	16	2, 11, 12, 30, 34
graduated cylinder, 50 mL	8	1, 8, 32, 34
graduated cylinder, 100 mL	8	4, 8, 16, 17, 19, 26, 31, 32, 35
hot plate	8	7, 8, 12, 34
laboratory apron	30	all
magnetic stirrer (if available)	8	23, 24, 25, 26, 31, 32, 35
micro-beaker	8	36
mortar and pestle	8	34
Petri dish, 11.5 cm diameter	8	28
pipet, 10 mL graduated	32	11, 12, 20, 24, 27, 33, 34
pipet, 25 mL graduated	8	33
pipet pump (or pipet bulb)	8	11, 12, 20, 24, 27, 33, 34
plastic tubing with two connectors	8	7, 10
power supply, direct-current	8	29
ring, 10 cm	8	16, 17
ring stand	16	2, 3, 4, 7, 8, 12, 13, 14, 15, 16, 17, 18, 19, 21, 22, 23, 24, 25, 26, 31, 32, 35, 36
rubber stopper assembly	8	7, 10
stirring bar	8	23, 24, 25, 26, 31, 32, 35
stirring rod	16	2, 4, 11, 12, 14, 16, 17, 18, 19, 21, 26, 30, 33, 34
stoppers, rubber-various sizes	40	8
syringe, 20 mL	8	6, 10
test leads	16	29
test tube, 13 x 100 mm	60	21
test tube, 20 x 150 mm	40	2, 3, 11, 12, 15, 20, 22, 34
test-tube rack	8	11, 12, 21, 34
thermometer	8	3, 15, 20
tongs, crucible	8	4
wash bottle	8	13, 21, 22, 23, 24, 25, 31, 32, 36

Consumables

Item	Amount	Experiment
Beral pipet	56	22
boiling chips	50	8
candle	8	17
cashew nuts	8	16
cola soft drinks, various	350 mL	35
cup, Styrofoam	8	1, 4, 18, 19
detergent	5 g	21
distilled water	30 L	1, 11, 13, 18, 19, 20, 21, 22, 23, 24, 25, 26, 27, 28, 29, 30, 31, 32, 33, 34, 35, 36
filter paper, 11.0 cm diameter	24	9, 28, 34
index card, 7.5 x 12.5 cm	8	17
lemon juice	400 mL	21
lemonade drink	400 mL	32
litmus paper, blue	20 pieces	21
litmus paper, red	20 pieces	21
marshmallows	8	16
multivitamin tablet with iron	8	34
paper towel	8	21
peanuts	8	16
matches	8 boxes	16, 17
popcorn	8	16
red cabbage juice	200 mL	21
rubber bands, small	40	9
salt	1 box	2
sand paper	1 sheet	28
soft drink	400 mL	21
swimming pool water	200 mL	33
tape, masking	2 rolls	9
tissue (preferable lint-free)	120	11, 13, 20, 30, 33
toothpick	8	36
tubing, bent-glass	8	8
vinegar	400 mL	21

vitamin C tablet, 500 mg regular	8	31
vitamin C tablet, 500 mg timed-release	8	31
weighing paper	8	19
wooden splints	16	16

Chemicals

Item	Amount	Experiment
acetic acid (17.4 M)	70 mL	13, 23, 27
aluminum chloride ($AlCl_3 \cdot 6H_2O$)	70 g	13, 14
ammonium hydroxide (14.8 M)	10 mL	23
ammonia, household	400 mL	21
barium hydroxide	1 g	26
benzoic acid	20 g	3, 15
boric acid	5 g	13
buffer solution, pH 7	500 mL	21
1-butanol	100 mL	9
calcium chloride	20 g	13, 14
citric acid	100 g	1, 32
copper, sheet	1	28
copper (II) sulfate pentahydrate	25 g	28
crystal violet	0.2 g	29
DPD free-chlorine powder pillows	50	33
ethanol, denatured (94–96%)	700 mL	8, 9, 10, 17
ethylene glycol	5 g	13
free-chlorine standard (50 mg/L)	100 mL	33
n-hexane	100 mL	9
hydrochloric acid (12 M)	300 mL	1, 13, 18, 19, 22, 23, 24, 34, 36
hydroxylamine hydrochloride	5 g	34
iron, sheet	1	28
iron (II) standard solution (100 g/L)	20 mL	34
iron (II) sulfate heptahydrate	30 g	28
iron (III) nitrate	15 g	20
lauric acid	150 g	15
lead, sheet	1	28, 29

lead nitrate	35 g	28
magnesium ribbon	5 m	1, 19
magnesium oxide	8 g	19
maleic acid	2 g	25
methanol	150 mL	9, 10, 13
nickel (II) sulfate	30 g	11
nitric acid (15.8 M)	30 mL	20
n-pentane	100 mL	9
1,10 phenanthroline	5 g	34
phenolphthalein	50 mL	23, 26, 36
phenyl salicylate	300 g	3
phosphoric acid (14.8 M)	5 mL	13
potassium nitrate	200 g	12
potassium thiocyanate	1 g	20
1-propanol	100 mL	9
silver, foil	1	28
silver nitrate	20 g	28
sodium bicarbonate	120 g	1, 21, 22
sodium bisulfite	10 g	22
sodium chloride	20 g	13, 14
sodium hydroxide	100 g	18, 21, 23, 24, 25, 30, 32, 35, 36
sodium nitrate	10 g	28
sodium nitrite	10 g	22
sulfuric acid (18 M)	70 mL	26, 29
zinc, sheet	1	28
zinc sulfate heptahydrate	30 g	28

Suppliers

Carolina Biological Supply Co.
1-800-334-5551
www.carolina.com

Hach Company
1-800-227-4224
www.hach.com

Frey Scientific
1-800-225-FREY
www.freyscientific.com

Sargent-Welch Scientific Co.
1-800-727-4368
www.sargentwelch.com

Fisher Science Education
1-800-955-1177
www.fisheredu.com

Science Kit and Boreal Labs
1-800-828-7777
www.sciencekit.com

Flinn Scientific Inc.
1-800-452-1261
www.flinnsci.com

Ward's Natural Science Establishment
1-800-962-2660
www.wardsci.com

Index

(by Experiment Number)